THE EGYPT AND PALESTINE CAMPAIGN

A Summary of the STRATEGY and TACTICS of THE EGYPT AND PALESTINE CAMPAIGN with Details of the 1917-18 Operations illustrating the PRINCIPLES OF WAR

TWELVE MAPS

BY

A. KEARSEY, D.S.O., O.B.E., p.s.c.

LATE LIEUTENANT-COLONEL, GENERAL STAFF

SECOND EDITION—REVISED

vide " SWISS MONTHLY MAGAZINE FOR OFFICERS "—
" This book offers a *good* summary of the events of the war on the Egyptian Front introducing the outcome of the strategic decisions for every year of the war."

vide " R.U.S.I. JOURNAL "—
" The comments are full and to the point."

CONTENTS

LIST OF MAPS.

GENERAL SUMMARY

OF THE

STRATEGY AND TACTICS OF THE CAMPAIGN.

On December 17th, 1914, Egypt, with the sanction of France, became a British Protectorate.

Turkish suzerainty disappeared. Our chief problem was to defend the main and vital artery of intercommunication of the Empire between the East and West, namely, the Suez Canal. Also, we made Egypt into a training ground for the Empire's reserve of Commonwealth and Indian troops. Thus, at first our policy was to secure Egypt and the hundred miles of canal between Suez and Port Said.

Later our policy changed, and led us into an offensive campaign against the Turks until, by the Armistice, we had reached Aleppo, 500 miles from the Canal. The reason for this was, in the first place, the need to economize troops by occupying the Qatiya Oasis, and thereby denying to the enemy a postion from which they could advance to the Canal and interfere with the shipping on it. If we held the whole length of the Canal a larger number of troops would be required than if we were astride the most likely line by which the enemy could advance towards it. Later, political motives caused us to occupy El Arish. Here we had superior numbers to any that the enemy could bring against us at the time. It was possible, therefore, to capture Rafa and Gaza before these places could be strengthened and while there was still an open flank round which we could use our great superiority in cavalry to the best advantage. Though we were checked at Gaza on March 26th and April 19th, 1917, our offensive operations enabled us to drive the Turks out of Egypt. This success was exploited until Jerusalem was captured by December 9th, 1917, and then until we entered Aleppo on October 26th, 1918.

Opposing Forces. The Turks hoped to be able to bring their Syrian striking force of 65,000 men against our Canal defences early in 1915.

In Syria, Djemal Pasha, the Turkish Commander, had the VIII Turkish Army Corps of three divisions, some troops of their XII Army Corps, with third line and depot troops in Syria. By the end of November, 1914, he had in Central Syria also the IV Corps, 10th Division, heavy artillery, pontoon and ammunition trains.

B

At Adana were left the III and V Army Corps. Their Hedjaz Division was brought up from Medina to Nekhl.

By the end of 1914 we had in Egypt the 10th and 11th Indian Divisions, the 42nd Territorial Division, Yeomanry, and Australian and New Zealand Divisions, bringing our total in Egypt up to 70,000, exclusive of 22,000 Egyptian and Sudanese troops.

Thus, we should have superior numbers to the Turks in the Canal area, although they were able during the war to enrol 2,700,000 men. In spite of poor organization and shortage of rations the Turks marched well, held positions stubbornly, and were quick and gallant in making counter-attacks. Their shooting was accurate with their guns, machine guns, and rifles.

The Germans supplied their Flying Corps and the personnel for their telegraphs and railways. In 1916 the Germans supplied Pasha groups as reinforcements. These consisted of a squadron, a battery, a battalion, and technical detachments.

Geography. The country of Egypt is a desert, except in the Nile delta and on the banks of the rivers, where it is fertile and habitable. The Suez Canal is approximately thirty miles from the habitable delta.

There were no metalled roads in the country except in the cities. The means of transport were by the railway along the western side of the Canal, and from Ismailia on the Canal to Cairo and to Alexandria. East of the Suez Canal is the Sinai Peninsula, approximately 120 miles broad. This was a formidable obstacle, in which water was scarce.

In the centre of the Sinai Peninsula is a waterless, stony plateau, rising in places to 3,000 feet. In the south the rocky hills rise to a height of 10,000 feet. In the north there is heavy sand. The water supply on each of these routes is precarious. The Wadi el Arish which runs between Nekhl and El Arish is dry for the greater part of the year. At times, between December and February after the rains, it is filled with water. During these times the rock cisterns throughout the peninsula become filled with water.

There are no roads fit for wheeled transport. For an advance on the Canal there were three possible routes. The southern route from Aqaba via Nekhl towards Suez and Kubri is over such rocky country as to be unlikely for any large force. The central route is from Beersheba or El Arish through Hassana and Jifjafa towards Ismailia. The northern route along the coast runs through Abd and Qatiya. For any large force the more likely line of advance would be towards Ismailia or Qantara. Further north the Suez Canal is protected by the Plain of Tina. The roughness and steepness of the country and the scarcity of water prevent any large bodies from crossing the southern route. The central route was the best and most sheltered road for an army advancing from the east. The northern route was vulnerable to shell fire from the sea. The Suez Canal is a formidable obstacle with both flanks secured by the sea. It is over 200 feet wide, with banks 40 feet high. There is an excellent field of fire from its eastern bank over the desert, except in a few places where there are sand-dunes giving cover south-east of Ismailia and east of the Bitter Lake and of Qantara. The Turks had no railway to help them across the Sinai Desert beyond Kossaima, and there was no route for mechanical transport. Their advanced bases were at Gaza and Beersheba. From these starting points the best line of advance for the Turks would be the central route across the desert, as it led to the sweet water canal at Ismailia. This canal could then be followed up to Cairo.

Communications. Behind their advanced bases the Turks had railways of different gauges. The main base of the Turkish Forces was at Haidar Pasha, opposite Constantinople, on the eastern side of the Bosphorus. From this base a single-line track of standard gauge ran through tò Muslimie, which is just north of Aleppo, and on to Rayak. This railway was incomplete through the Taurus (11,000 feet) and Amanus (6,000 feet) Mountains.

All troops and transport had to cross these two ranges by road in the early stages of the campagin. The tunnel through the Taurus Mountains was completed only just before the Armistice. The metre gauge line was continued from Mus-limie through Aleppo to Rayak. Then for 250 miles to Southern Palestine there was great difficulty in maintaining the railway service owing to lack of fuel and repair shops and material. From Rayak the single track metre gauge ran west to Beirut and south through Damascus to Deraa, where it branched in a westerly direction through Afule to Haifa, and ran south via Maan to Medina, 250 miles from Mecca. IA French line of railway ran from Jaffa to Jerusalem. It took as long as six weeks for reinforcements to arrive from Haidar Pasha to the Turks' Palestine front at Gaza and Beersheba. The distance was approximately 1,275 miles.

Plans for 1914. Owing to the difficulties of transport and movement, the Turks, out of their striking force, were only able to organize a force of approximately 12,000 men for their raid on Egypt.

Djemal Pasha began his preparations in November for a raid on Egypt, where he hoped to add to our difficulties by raising a revolt in favour of Turkey, and he wished to interrupt the traffic on the Suez Canal. He decided to advance by the three routes across the Sinai Peninsula, and to make his main attack by the central road against our positions at Deversoir, Serapeum, and Tussum. It was in this area that the Turks could hope to obtain quickly the much-needed fresh water if they were successful in crossing the Suez Canal.

Our plans were to use the Suez Canal as an obstacle. The 10th and 11th Indian Divisions with the Imperial Service Cavalry Brigade and the Bikanir Camel Corps were responsible for the Canal defence, helped by the fire of the ships from the Canal. The 10th Division and part of the 11th Division provided the Canal posts.

A reserve was retained at Moascar, near Ismailia.

Operations in 1914. By November 15th the Turks had approximately 8,000 men in El Arish. By November 20th the Turks were in force twenty miles east north-east of Qantara in touch with our Bikanir Camel Corps. By December 31st our forces in Egypt numbered 70,000.

The Turks had by this date formed advanced depots at Khan Yunus, El Arish, Auja, and Kossaima.

Operations in 1915. By January 26th our forward troops were in touch with the Turks near Qantara. Four cruisers occupied their stations in the Canal to support the troops at Qantara, Ballah, Shatt, and Shallufa. By February 2nd the situation was that there were approximately 10,000 Turks with guns and pontoons east of Serapeum and Ferry Post. It was estimated that they had nine batteries and two six-inch howitzers in the vicinity of the Canal.

Two thousand Turks supported by guns were east of Qantara. On the night of February 2nd/3rd the Turks advanced to attack our posts at Qantara, Ferdan, Ismailia, and Tussum. Their attempts to cross the Canal between Serapeum and Tussum completely failed. Only three pontoons got across, and all the occupants were either killed or captured. Their artillery was effectively dealt with by the guns of the *Hardinge, Requin,* and *d'Entrecasteaux.* From Serapeum we made a successful local counter-attack on the east bank of the Canal. The main result for the Turks of this raid was that the night shipping in the Canal was suspended for a week, and day shipping was stopped for twenty-four hours. At 1400 hours on February 3rd the Turkish artillery ceased fire, and shortly afterwards the whole force began to withdraw by the routes by which they had advanced. Their casualties were over 2,000. Our pursuit was not vigorous. On February 4th the Imperial Service Cavalry Brigade crossed at Ferry Post and captured a small convoy, but did not closely engage the enemy. By the evening of February 4th the yeomanry in

Cairo were sent by train to Ismailia, but there were no means of rapidly transporting so many mounted men across the Canal, nor was the water supply organized for operations in the desert with the available camel, mule and donkey transport. The result was that the Turks were able to retire with all their guns and baggage.

This raid caused us to reconsider plans for the defence of the Suez Canal. We now realized the extent to which a descent on Egypt from the East was practicable, and the fact that a purely passive defence would enable the Turks to interfere with the traffic on the Canal, through which it was essential to maintain open intercommunication between the East and our main theatre of war on the Western Front.

Our plans now, therefore, were to make dispositions for a more active defence of Egypt, while the Turks were fully occupied with expeditions up the River Tigris and in Gallipoli.

The Turks had no available troops for another raid during 1915 across the Sinai Peninsula, and, consequently, they made efforts to stir up Senussi hostility in the Western Desert. In this area there was much unrest among the Arabs. It was, therefore, found necessary to withdraw our posts in the Western Desert to Mersa Matruh, and to concentrate there a sufficient force to secure the Alexandria-Dabaa railway, to deal with local situations and to reconnoitre the Maghara Oasis. This force, consisting of a yeomanry brigade and an infantry brigade, was formed by November 20th. Early in December the Senussi began to assemble in this area. By December 20th our composite brigade was assembled at Mersa Matruh. It consisted of a yeomanry regiment, a section of horse artillery, the 15th Sikhs less two companies, and six armoured cars. The Senussi in the vicinity of the Wadi Senab, Sahifa, and Ras Manaa were estimated at approximately 1,000 to 1,500 and two guns. Between December 11th and 13th they were driven out of the Wadi Senab.

It was fortunate that the Senussi adopted a regular form of war, and were prepared to make a stand and to hold positions instead of adopting the usual methods of guerrilla warfare. Had they scattered and harried our lines of communication, and made full use of their mobility to operate between the different oases on the Libyan Desert, our difficulties in dealing with them would have been much greater.

After the Wadi Senab had been cleared, the Senussi collected in the Wadi Majid. They occupied a commanding position on Gebal Medwa Hill, six miles south-west of Mersa Matruh. This position was attacked by two columns from different directions. The enemy was driven out of his position with considerable loss on December 25th.

Operations, 1916. The operations of 1916 started with further fighting against the Senussi. On January 23rd two of our columns left their camp at Bir Shola, twelve miles south-west of Mersa Matruh. One column composed of infantry and a small proportion of cavalry and guns, made a frontal attack, while the other column, containing six squadrons, a horse artillery battery, and a machine-gun section, operated against the enemy's southern flank at Halazin. The Senussi were driven from their entrenched camp by the infantry by 1445 hours. The cavalry, however, were too much exhausted owing to lack of water to carry out a serious pursuit.

The result was that the bulk of the enemy was able to withdraw. By February 20th, 1,600 Senussi with four guns were located at Agagiya, sixty-five miles west of Mersa Matruh. By February 23rd our column, consisting of four squadrons of yeomanry, three battalions, and a horse artillery battery, reached a position at the mouth of the Wadi Mehtila, eight miles north-east of the enemy's position. Early on February 26th the yeomanry moved out of camp and accurately located the enemy's position. The infantry made a frontal attack, and drove the Senussi out of their position with the close co-operation of the artillery and machine guns.

The Yeomanry remained in a position of readiness out of sight of the retreating Senussi until the ground was suitable for their mounted action. Their opportunity arrived about 1400 hours, when the order was given to charge.

The enemy were completely surprised and routed. Their leader, Jafar Pasha, was captured, and they were never again able to fight as an organized force. Sollum was reoccupied on March 14th. On March 19th an armoured car detachment, under the Duke of Westminster, started across the desert to find the imprisoned crew of H.M.S. *Tara*. This detachment covered 240 miles in twenty-four hours, and was successful in bringing back forty survivors of the crew.

Our plans for 1916 were to maintain an active defence in Egypt and to reduce the number of troops required for the security of the Suez Canal. The first step was to be the occupation of the Qatiya area, where the railway and pipe-line were to be brought for the maintenance of troops in the forward area. The Qatiya Oasis was to be held with one division and three mounted brigades. A reserve of three divisions was to be held on the Canal. The thirteen divisions in Egypt were to be sent to other theatres of operations as soon as possible.

Actually, before the end of March, six divisions had left Egypt. By denying to the enemy the Qatiya-Romani area, a suitable assembly place for the Turks, it would be possible to organize a mobile column for offensive operations, and would enable us to abandon the long and elaborately fortified line east of the Suez Canal. It would also be possible to keep the enemy from bombarding the Canal with long range artillery fire. The enemy would have the disadvantage of crossing the sixty miles of sandy desert from El Arish if they were to attempt to assume the offensive in the area in which their chances of success were most probable, as the water-bearing prospects were far greater and more suitable for a large force than in any other quarter in the vicinity of the Canal. For the Turks advancing on the Canal the country between Kossaima and El Arish was important. These were the two nearest road centres for a force advancing from the east towards the Canal. There should be no surprise attack on troops defending the Canal if these areas were adequately watched by mobile troops and R.A.F. It was estimated that of the 250,000 Turkish troops in Syria, a force of approximately 40,000 could be maintained with water in the Wadi el Arish, and a similar number in the Qatiya-Romani area, Neither of the other two routes leading to the Canal possessed such favourable lines of advance or such an abundant water supply. For the supply of the troops in the Romani area a railway of standard gauge and a pipe-line of Nile water were brought out from Qantara. Steps were also taken to add to the mobility of the force by organizing camel companies, each of 2,000 camels, and also donkey companies, each of 2,000 donkeys. Finally, 35,000 camels and 8,000 donkeys were organized on a company basis. Trains also were organized, namely, six divisional trains, each of seventy-two limbered G.S. wagons, and two mixed horse and motor transport trains, each capable of carrying seventy-two tons of supplies. Owing to our organization and forward policy the initiative passed to us, and, in spite of two advances made by the Turks during 1916, we continued our advance through Palestine and Syria to Aleppo. On March 10th, General Sir John Maxwell handed over his command to General Sir Archibald Murray. The force in Egypt now became the Egyptian Expeditionary Force.

A point for consideration now was whether the Turkish operations towards Baghdad and towards the Canal might not be dealt with by one amphibious expedition from Alexandretta based on Cyprus. An advance in force towards Aleppo would then enable us to be astride the Turks' communications, and would save our long advance up the Tigris and up through

Palestine and Syria with large forces. Comparatively small forces could safeguard our interests in the Basra Vilayet and in the Qatiya Oasis. There would be considerable economy of force on our part, and great saving in man-power, material, and in resources.

We had failed in our Gallipoli campaign. The great difficulties of amphibious warfare had been clearly brought home to us, and though we had superiority of sea power with surface ships there was the increasing and costly menace of the submarine. In addition, the troops available for this expedition to Aleppo would have been those released from Gallipoli, and though there were now in Egypt twelve infantry divisions, two brigades of Indian infantry, dismounted yeomanry, and mounted brigades, yet they required a rest before undertaking further active operations.

It was not possible to calculate how many could be taken for this operation, as reinforcements might be required either for the Tigris operations or for the Salonika campaign.

Our landing, too, might be opposed by the bulk of the 250,000 Turks reported to be in Syria and Palestine. The scheme for the landing at Alexandretta was, therefore, abandoned.

By the middle of April Kress von Kressenstein undertook an enterprise along the northern route. On April 23rd his force, estimated at 3,650 men, six guns, four machine guns, and six companies of camelry, successfully attacked our advanced posts at Qatiya and at Oghratina and Hamisah.

Their attacks on our posts at Dueidar and Hill 70 were unsuccessful. In these engagements the 5th Mounted Brigade was protecting the railhead five miles west of Romani with the following detachments: A squadron, a machine-gun detachment and dismounted details at Qatiya. At Oghratina there were two squadrons less one troop and an R.E. detachment. At Hamisah were three squadrons and one troop. The headquarters and remainder of the 5th Mounted Brigade were at Romani. Dueidar was occupied by 156 rifles and Hill 70 by one battalion. At 0415 hours the post at Oghratina was first attacked from three sides by the Turks, who were then not more than fifty yards from our forward defences. The defenders, after two and a half hours of stubborn resistance, were forced to surrender. The Turks then attacked the Qatiya force, which, though reinforced by a squadron of the Worcesters from Hamisah, was unable to hold out after 1500 hours. Early on April 23rd, the Commander of the 5th Mounted Brigade with the Warwickshire Yeomanry had made a raid on the enemy's camp at Mageibra. This he found empty, as the occupants of it had gone to raid our posts at

Dueidar and Hill 70. The commander of the five troops and machine-gun section who had been left at Romani marched out to relieve the Qatiya garrison, but there was no co-operation with the Brigade Commander's force and little could be effected. Our troops at Dueidar successfully maintained their position.

They were not surprised at any of the six works surrounding the Dueidar Oasis. The Turks had heavy casualties throughout the day.

They were pursued by the 5th Australian Light Horse Brigade and by the Flying Corps.

After this Romani was occupied by infantry and the railhead was advanced. A line of fortifications, consisting of eighteen infantry redoubts, each with 100 rifles and two machine guns, was constructed between Mahamdiyeh and Katib Gannet covering Romani. The Anzac Mounted Division watched this front and made many long-distance reconnaissances.

During the summer Senussi forces under Sayyid Ahmed, operating from the oases of Bahariya, Farafra, Kharga, and Dakhla became troublesome.

The 53rd Division and some brigades of dismounted yeomanry were distributed between Faiyum (eighty miles south-west of Cairo) and Assuan. It was not until February, 1917, that Sayyid Ahmed was finally defeated, and his forces were dissipated by a detachment of armoured cars from Sollum. Ali Dinar, of Darfur, who had also been hostile, was defeated near El Fasher by a force under Colonel Kelly, commanding Egyptian troops, during May. Ali Dinar escaped, but he was killed on November 5th, when his following of 2,000 men was completely routed. The Sherif of Mecca led the Arabs of the Hedjaz against the Turks. Jeddah was captured on June 12th, and Taif fell on September 23rd. Medina in the meantime was invested. During the summer the Arab forces in front of Medina were joined by Colonel Lawrence. He persuaded Feisal, the Arab Commander, to withdraw his forces to Wejh and to menace the Turkish communications. This caused the Turks to dissipate their forces along the Hedjaz railway. At Wejh an Arab training depot was formed. Colonel Lawrence then led a force of Arabs against Aqaba from the north and captured it on July 6th.

This place then became Feisal's base for the Arab force. Later, this Arab force protected the right flank of the British army in its advance through Palestine, and, finally, closely co-operated in the defeat of the Turks in the fighting between Damascus and Aleppo in 1918.

After the advance to Qatiya in April, Kress von Kressenstein did not make any further movement until July. On July 19th our R.F.C. reported that approximately 8,000 Turks were in the vicinity of Abd, Bir Jemeil, and Bir Bayud. On the following day a large force was reported to be at Oghratina, and that 3,000 Turks of the 3rd Turkish Division with eight machine-gun companies were at Mageira. On August 2nd the Turks were located between Mahamdiyeh, Katib Gannet Hill, and Etmaler. Their force was estimated to be fifteen battalions, two batteries of mountain artillery, heavy artillery (making a total of thirty guns), engineers, thirty-eight machine guns, and two camel corps companies. Their force was estimated at 18,000 when it left Shellal.

On the evening of August 3rd the main body of Kress von Kressenstein's force followed up the 2nd Australian Light Horse Brigade as it withdrew back to its position on Wellington Ridge. At this time our left flank rested on the sea, and our right flank on Etmaler Hill and Katib Gannet, covered by troops at Hod el Enna, Mount Meredith, and Wellington Ridge. The Turks meant to envelop our southern flank and to gain a position on the railway between Qantara and Romani. At midnight on August 3rd/4th they pressed their attack against the outposts on our right flank, held by the 1st A.L.H. Brigade from Katib Gannet to Hod el Enna. The Turks were in greatly superior strength, and were able to drive back the 1st A.L.H. Brigade over Wellington Ridge to Mont Royston. Early on August 4th they made, in addition, frontal attacks, supported by heavy artillery from Abu Hamra, against our entrenched position running south from a point a mile east of Mahamdiyeh Station.

Our right flank extended to a point beyond which they made their enveloping attack, and, consequently, in attempting to prolong their line of attacking troops in order to overlap our flank their two forces lost touch. Their southern force, intended to envelop our flank, became involved in heavy sand hills one and a half miles from the railway by the early afternoon on August 4th. Two mounted brigades were sent to make a counter-attack, and by 1830 hours had regained Mount Royston.

The Turks now made no effort with their reserves, or by pressing their frontal attacks to restore the situation. Instead, when their enveloping movement failed they fell back in disorder. We still had in reserve five battalions in the vicinity of Etmaler, and three brigades posted along the railway between Qantara and Pelusum. Also from No. 2 section were Smith's mobile column and the 3rd A.L.H. Brigade available for pursuit. However, although we recaptured Wellington Ridge at daybreak on August 5th, yet we could not effectually follow up the Turks, who were adequately covered by rearguards, owing

to the difficulty of co-ordinating the movements of the
separated brigades of the mounted division, and owing to the
difficulties of water supply and the intense heat, which was
most exhausting for infantry marching in soft sand. The
Turks were able to withdraw with a loss of approximately
5,000 casualties, four guns, nine machine guns, and 4,000
prisoners.

On August 7th they were on the line Hod el Masia-
Oghratina. On August 9th they were able to repulse a direct
attack by the Anzac Mounted Division in the vicinity of Bir el
Abd. They then withdrew towards El Arish. The Turks
undertook no further offensive actions, and they were
gradually driven into Palestine. General Murray now decided
to clear the Sinai Peninsula and to occupy El Arish. Accordingly he formed an Eastern Force for this advance. This
force was under the command of General Dobell, and consisted
of all the troops east of the Canal.

General Dobell further split up the forces at his disposal by
forming a Desert Column as a striking force. This column
consisted of the 42nd, 52nd, and Australian and New Zealand
Divisions, under General Sir Philip Chetwode. The troops
were to be accompanied by the standard railway and the pipe-
line from Qantara. The rate of advance for the railway was
about 20 miles a month. By December 9th the Turks at
El Arish were estimated to be 1,600 infantry in strongly
entrenched positions.

During the night December 20th/21st the Australian and
New Zealand mounted troops and Camel Corps marched on
El Arish, supported by the 52nd Division. Soon after daybreak on December 21st mounted patrols found that El Arish
was unoccupied. The R.F.C. reported that 1,600 Turks were
marching towards Magdhaba and Abu Aweigle, and that
Maghara had been vacated. Our troops then occupied El
Arish and stores were landed on the beach. The Turks had
prepared a system of five well-constructed redoubts at
Magdhaba on both banks of the Wadi el Arish. By daybreak
on December 23rd the mounted division of the Desert Column
and the Camel Corps Brigade were within four miles of these
redoubts. Converging attacks were simultaneously made on
the Turks' defences, and were pressed with the greatest determination for eight hours.

When No. 1 Redoubt had been captured by the combined
attack of the 3rd Australian Light Horse Regiment working
up the Wadi with two companies of the Camel Corps Brigade
advancing dismounted across the open from the north-west,
the whole defence collapsed. The Turkish Commander and
two battalion commanders with 1,280 prisoners were captured.

1917. Our policy was still the active defence of Egypt. Our plan was to continue to advance by the coastal route, and to capture Rafa and then Gaza on the same principle as had already been successfully carried out in the capture of Magdhaba.

Full use was to be made of the factors of mobility and surprise. By the capture of Gaza the Turks would be forced to evacuate Southern Palestine, as we should be behind their base at Beersheba. These operations were to be carried out as soon as forward supply depots could be organized and replenished, and camel and horse-transport trains could be arranged.

We could confidently expect to have superior numbers at the decisive time and place. The enemy in Southern Palestine could not have more than 1,500 sabres, 12,000 to 16,000 rifles, and 74 guns. Our own troops would be the 52nd, 53rd, and 54th Divisions, the Australian and New Zealand Mounted Division, and the Imperial Mounted Division, a light-car patrol and two light-armoured motor batteries, making a total of 8,500 sabres, 92 guns, and 25,000 rifles. There was also the 74th Division being formed in Egypt from yeomanry units. In addition, 55,000 men in the Egyptian Labour Corps were formed. The plan which the Turks had arranged was to keep the bulk of their force in a central position south of the Wadi Hesi with detachments at Gaza, Sheria, Hureira, and Beersheba.

In order to carry out our plan it was necessary to increase the mobility of the Eastern Force and to reorganize the supply and transport arrangements, as the railway had only just reached the Wadi el Arish. Therefore, seven camel trains, each with fifty-five camels, were organized as first line transport. Each train was capable of carrying seventy-two tons. Also six divisional trains were organized, each with seventy-two limbered G.S. wagons, while two mixed horse- and motor-transport trains, each capable of taking seventy tons, were also organized.

If we could advance rapidly on Rafa and then on Gaza before the Turks could withdraw from Beersheba and Wadi el Arish their communications would be in danger, as we should be behind their advanced base. Since their defeat at Romani the Turks had not attempted to take the initiative against us. It was advisable, therefore, to continue to press our offensive and to gain such an important point as the fortress of Gaza while their morale was low, and while we still had greater numbers for an attack against it than they could collect locally. By operating in the coastal area we gained the advantage of being in country suitable for mounted action. Our superiority in mounted men was very considerable. In addition, we should be helped by sea-power with the ships securing our left flank. Again, in this Plain of Philistia north of Gaza there was one road to Gaza from Jaffa and Ramleh, and another from Jerusalem via Latrun.

There was ample water from the wells during and after the regular rainfall from November to March. Crops were plentiful between April and June.

The first operation in 1917 was the capture of the Turks reported to be holding four strong works at Magruntein, 2,000 yards south-west of Rafa. Early on January 9th the Turks in these entrenched works were surprised by the Australian and New Zealand mounted troops and the Imperial Camel Corps, who had made a twenty-nine-mile night march. The N.Z.M.R. closed the exits from the north. The 3rd and 1st A.L.H. Brigades attacked the Turkish works facing east, the Camel Corps Brigade attacked the works facing south-east, while the 5th Mounted Brigade operated against the redoubt on the south-west. The Turks were not disposed in depth and had no reserves. Each of their works was isolated and self-contained. Our troops were in touch with each other right round their defences, and all had definite objectives. The Turks, however, defended stubbornly, and at 1630 hours, as we had not been able to gain our objectives, General Chetwode ordered a withdrawal as information was received that Turkish reinforcements 500 strong were advancing from the north and north-east.

The determination of the troops, however, changed the situation. The N.Z.M.R. Brigade, before the order to withdraw reached them, had resolutely pressed their attack from the north against the Turks' northern redoubt. This was captured after a bayonet charge. Then the 1st and 3rd Australian Light Horse Brigades charged, but the Turks surrendered before these attacks were pressed home. One thousand six hundred prisoners, four guns, and six machine guns were captured.

It was now reported that the Turks were concentrating near Shellal, and that 5,000 were being sent to the vicinity of Gaza. Also they had made strong works at Weli Sheikh Nuran.

By the end of February our mounted troops reached Khan Yunus. It was now definitely decided to capture Gaza as soon as the railway reached Rafa. All supply arrangements were made to enable troops to carry through the operation in one day. If Gaza did not fall within this prescribed time limit, it would be necessary to withdraw back to the Wadi Ghazze for food, water, and ammunition. Success before nightfall on the day of attack was thus an indispensable condition. If Gaza was captured within twenty-four hours of the commencement of hostilities there was ample water there for all men and animals, and supplies could be brought to the coast in the vicinity of this town by sea.

By March 5th the Turks began to fall back on Gaza, Sheria, and Beersheba. By March 16th the broad gauge railway reached Rafa. By this date, also, the Desert Column was in the vicinity of Shaikh Zowaiid. By March 25th our force was assembled and ready to move on Gaza. Advanced Headquarters were at El Arish. The Desert Column was at Belah, the 54th Division was at In Seirat, the 52nd Division was at Khan Yunus, and the Camel Corps and armoured batteries were about Abassan el Kebir. On March 26th the first battle of Gaza was fought.

The enemy were estimated to have in the vicinity 1,500 sabres, 16,000 rifles, and 74 guns.

Our strength was 8,500 sabres, 25,000 rifles, and 92 guns. Actually at Gaza it was estimated that there were 4,000 rifles and 20 guns. Our plan of attack was to send the Mounted Divisions and Camel Brigade across the Wadi Ghazze at Sheikh Nebhan, advance to Mendur, and then to form a screen to the north-east and east of Gaza, to hold off enemy reinforcements and to prevent the garrison of Gaza from escaping in these directions. The 53rd Division and one infantry and one artillery brigade of the 54th Division were to assault Gaza from Ali Muntar to Samson Ridge.

A small detachment of one battalion of infantry and two squadrons of yeomanry was to advance up the coast towards Shaikh Ajlin to cover the left flank. These troops for the assault on Gaza were under the command of the G.O.C. Desert Column.

The remainder of the 54th Division was to form a support to the mounted troops by occupying a position at Mansura and Shaikh Abbas. The 52nd Division was to form the general reserve west of El Breij and to protect the line of communications. The headquarters both of the Eastern Force and of the Desert Column were south of the Wadi Ghazze at In Seirat. On March 26th at 0230 hours the A. and N.Z. Division left its bivouacs and reached Beit Durdis by 1000 hours. It was followed by the Imperial Mounted Division, who reached Mendur at 0930 hours. The Imperial Camel Corps also went to Mendur.

The 53rd Division started at 0100 hours. By 0745 hours the 158th Brigade was on Mansura Ridge.

When the fog lifted at 0820 hours the Turks were completely surprised to find our leading brigade within 4,000 yards of their position. But, unfortunately, two hours had been lost by the 53rd Division in waiting for the fog to clear, and the attack on the Turks' trenches was not launched until nearly midday. Had the infantry been close to the Turkish position, as they could have been had they not been delayed by the fog, the attack might have started earlier, and a decision might have been obtained before nightfall. When the infantry attack did start the enemy had recovered from their surprise, and the advance had to be made over the open plain from Mansura by the 158th Brigade towards Ali Muntar and up the exposed Sire Ridge by the 160th Brigade. The reserve brigade, the 159th, was early sent in to attack on the right of the 158th Brigade.

It was not until 1600 hours that the 161st Brigade of the 54th Division co-operated in the attack. This was due to the fact that close touch was not maintained between the 53rd Division and the 161st Brigade, and the order to this brigade took a long time in transit. However, by 1800 hours Green Hill had been captured, and the whole Turkish position on the Ali Muntar Ridge was in our possession.

The 2nd A.L.H. Brigade and the New Zealand Mounted Brigade were on the northern outskirts of the town, and were in touch with the infantry on Ali Muntar.

Now the decision had to be made whether to push in to Gaza or not. Unless Gaza was captured and the water in it obtained for the mounted troops, it would be necessary for these troops to withdraw. If they went back to water they would leave the infantry in their present positions with their eastern flank exposed, and a considerable gap between the two brigades of the 54th Division at Shaikh Abbas and the troops on Ali Muntar. There was a risk in continuing the attack just as it was getting dark, as there was increasing pressure by Turkish reinforcements converging on Gaza. Some 7,000 Turks with a cavalry screen were reported to be advancing from Hureira; from Deir Sineid and Huj also large bodies were reported to be approaching. Soon after 1800 hours General Dobell issued orders for the mounted troops to withdraw to the Wadi Ghazze. Previously he had ordered the 54th Division to move in to the right of the 53rd Division in order to close the gap between the two divisions. This order was not known to the Commander of the attacking troops, who ordered the Commander of the 53rd Division to withdraw his right back in order to gain touch with the 54th Division. Thus the ground dominating Gaza from the south-east, for which 4,000 casualties had been incurred, was abandoned. The result of these orders was that on the morning of the 27th the 53rd Division was in a position facing west through Sheluf, while the 54th Division was in a position facing east through Mansura and Burjabye. The Turks could then occupy the vacated dominating positions of Ali Muntar and Shaikh Abbas. By 0930 hours on March 27th the 53rd and 54th Divisions were back to back with not more than two miles of intervening ground between them. The attempts by the 160th and 161st Brigades to occupy Ali Muntar had failed. Their leading troops had gained a position on Ali Muntar and Green Hill, but had later been unable to retain them owing to heavy Turkish counter-attacks. There was now no longer any hope of capturing Gaza, and the present position of the infantry in the salient was untenable. Orders were, therefore, given for a withdrawal to the west bank of the Wadi Ghazze. It is summed up in despatches as follows: " If it had been practicable for the G.O.C. Eastern Force to advance with his three infantry divisions and two cavalry divisions I have no doubt Gaza would have been taken and the Turks forced to retire, but the reorganization of the force for a deliberate attack would have taken a considerable time; the horses of the cavalry were very fatigued, and the distance of our railhead from the front line put the immediate maintenance of such a force with supplies, water, and ammunition entirely out of the question. The only alternative, therefore, was to retire the infantry." Later the Commander-in-Chief wrote: " The operation was most successful, and owing to the fog and waterless nature of the country round Gaza just fell short of a complete disaster to the enemy."

As indicated in despatches, victory had been very nearly gained. Actually, the Turks were on the point of abandoning their position. The risks, however, of continuing the attack were considered to outweigh the possible chances of success. We captured 837 prisoners and two guns at a cost of 3,960 casualties.

By March 29th a defensive position was organized west of the Wadi Ghazze, and was held by the 54th, 52nd, and 53rd Divisions. The Turks strengthened and wired their defences up to Atawineh Ridge. By April 5th our railhead was at Belah. Supplies were landed at this place on the open beach.

Our next attempt to capture Gaza was to be in two stages. The first stage was an advance to the line Shaikh Abbas, Mansura, and Kurd Hill. These objectives were captured early on April 17th with little difficulty.

These positions were consolidated and tanks and heavy artillery were brought up.

The second stage was started on April 19th. The plan was to subject the Turkish trenches to a severe bombardment, and then to deliver frontal attacks from Birket es Sana to the sea.

The 53rd Division was to advance up Samson Ridge towards Shaikh Ajlin. The 54th and 52nd Divisions, under the command of the G.O.C. 52nd Division, were to capture the Khirbet Sihan works and also Ali Muntar and Outpost and Middlesex Hills, south of Gaza. The Camel Corps Brigade was to operate on the right flank of the 54th Division. On the right of the Camel Corps Brigade the mounted divisions were to attack, dismounted, the Atawineh trenches and seize the Birket es Sana spur and demonstrate against the Hureira Redoubt. Our preliminary bombardment with gas shells produced little result on the Turkish trenches, guns, or Machine guns on the large front which was to be attacked. The few tanks available were unable to reach the Turks' main positions. The limit of our advance was Samson Ridge on our left by the 53rd Division; Outpost Hill was captured by the 153rd Brigade; further to the east the left of the 54th Division reached the Turks' forward trenches, but were then so enfiladed by fire from Ali Muntar that they could not advance. The 161st Brigade reached Tank Redoubt, and the Camel Corps Brigade penetrated to Khirbet Sihan. The Turks, however, had strong positions on their 15,000 yards frontage, and our numbers were not so superior to theirs as to warrant success against an enemy who was always staunch in defence, and who in this case had all the advantage of position and field of fire and observation against an enemy advancing from the south. We had 11,000 sabres, 24,000 rifles and 170 guns. The enemy had 1,500 sabres, 18,000 rifles, and 101 guns.

By 1500 hours our casualties had been 6,400, our advance was definitely checked, and the enemy still had their reserves in hand. At 1820 hours we were forced by Turkish counter-attacks to evacuate Outpost Hill. Our line then ran from the sea at Shaikh Ajlin through Samson Ridge, south of Outpost Hill, through Shaikh Abbas, Meshrefe, to the Wadi Ghazze. One brigade of the 74th Division was at Mansura ready to support the 52nd Division.

To deploy any more troops in further frontal attacks would only lead to casualties. Artillery ammunition supply was becoming short. Orders were therefore issued that the action should be broken off, that all ground occupied should be consolidated, and that troops should be prepared to resume the offensive after a bombardment at dawn on the 20th. Turkish counter-attacks were repulsed. The opinion of the Commander of the Desert Column was that " in view of the great strength of the positions to which he was opposed, the renewal of a direct attack with the force at his disposal would not be justified by any reasonable prospect of success."

Plans for further offensive operations were abandoned until

c

reinforcements arrived from Salonika. There was position warfare now in front of our position from the Wadi Ghazze at Tel Jemmi on our right flank through Sharta and Dumb Bell Hill to Shaikh Abbas. Our line made an angle at this point, and ran in a north-westerly direction by Mansura, Blazed Hill, Heart Hill, Samson Ridge, to the sea at Shaikh Ajlin.

On May 23rd the railway extension between Beersheba and Auja was destroyed over a distance of 13 miles. Reinforcements began to arrive during June. The 10th and 60th Divisions and the 7th and 8th Mounted Brigades came from Salonika. Units from India and Aden with those in Egypt enabled the 75th Divisions to be formed. French and Italian contingents were also added to the Egyptian Expeditionary Force. Communications were improved, and a branch railway was made from Rafa to Shellal. The pipe-line was brought up to Belah. Reservoirs were built at Belah, Tel Jemmi, Sitta, Khan Yunus, Shellal, Rafa, and Qantara. The Cabinet decided to continue operations for the invasion of Palestine. To carry out these operations, General Sir Edmund Allenby was appointed to command the Egyptian Expeditionary Force. He assumed command of the Egyptian Expeditionary Force on June 28th. From this date until the Armistice in 1918 this campaign becomes particularly interesting, not only because it was eminently successful, but because it was most carefully planned and vigorously executed. Full use was made of our great numerical superiority and of our enormous resources to carry out an offensive campaign. Successes were followed up and turned into complete victories, so that the campaign became a factor in deciding the World War. There was reasonable prudence in the conduct of operations, combined with the greatest determination in carrying out plans to the limit of the endurance of the troops and in accordance with the possibilities of transport and supply. Periods of rest and reorganization were followed by offensive action, energetically continued until the objectives were reached. Difficulties of country varying from sandy plains and enclosed river-valleys to steep, rocky, wind-swept hills were overcome. The climate, varying from cold rain in the hills to choking khamsins in the plains, was stubbornly endured. The intricate problems in the field which the Commander had to decide were dealt with so firmly, so promptly, and so effectively, that the co-operation of all arms was assured. This confidence and unity led not only to an unqualified success, but showed how military operations in mobile warfare can and should be conducted.

Between June and October, while the Egyptian Expeditionary Force was being reinforced and reorganized, the Turks also were improving their communications, and adding to the strength of their garrisons and their works. They constructed a chain of localities along their thirty-two-mile front between Gaza and Beersheba. Besides the works immediately covering these two places, they had made redoubts at Kauwukah, Hureira, Baha, Atawineh, and Sihan. Only between Kauwukah and Beersheba, four and a half miles apart, was there such a gap that there was no close support between the defences. They had a good metalled road connecting Gaza with Beersheba. A 3 ft. 6 in. gauge railway ran to the Wadi Hesi, north of Gaza, and to Beersheba through Sheria. General Allenby now organized the Egyptian Expeditionary Force into three corps: the XX Corps, consisting of the 10th, 53rd, 60th, and 74th Divisions; the XXI Corps, consisting of the 52nd, 54th, 75th Divisions and a composite force; and the Desert Mounted Corps, consisting of the Anzac, Australian, and Yeomanry Mounted Divisions. Our strength was 18,000 sabres, 80,000 rifles, and 450 guns. The Turks had two corps and a cavalry division, namely, 1,500 sabres, 50,000 rifles, and 300 guns. The courses open to General Allenby for his offensive operations were governed by the following considerations. Gaza was a very strong position guarding the coastal route, which was the most favourable ground for his operations, with the left flank secured by the navy.

The country between the Wadi Ghazze and Beersheba was waterless, and there were no metalled roads available south of the main Gaza-Beersheba road. There would be wide separation between two forces operating at Gaza and Beersheba. There would be great difficulty in surprising the enemy by an attack on Beersheba owing to the great quantity of transport required for supplies and water; and there would be great difficulty in providing the necessary transport for a large striking force in the vicinity of Beersheba.

A direct attack on Gaza might lead to no decisive results, as the enemy could fall back on the wadis north of Gaza at right angles to our line of advance. Similar disadvantages occurred if the Atawineh position was penetrated.

The available transport, namely, 30,000 camels. was sufficient only for the supply of a corps with food and ammunition up to Beersheba and one more march, but it would be entirely dependent on the water supply there. Therefore, the capture of Beersheba on the day of the assault was essential.

The plan for the attack of the Turks on the Gaza-Beersheba line was that the enemy's left flank was to be turned by the Desert Mounted Corps' attack with two divisions, and the 7th Mounted Brigade from the east and north-east. Two divisions of infantry were to attack the main Beersheba defences from the south-west, while one division covered the left flank of the attacking force. The XXI Corps was left for the attack on Gaza. One mounted division was to hold the centre. Beersheba was to be captured as rapidly and secretly as possible, then the Turks' left flank defences at Hureira and Sheria were to be captured and their garrisons pushed back on Gaza, while the cavalry advanced towards the water supplies in the Wadi Hesi, and cut off their retreat from Gaza. In order to mystify and mislead the enemy as to our actual point of attack, the bombardment of Gaza was begun on October 26th and was continued steadily for eight days and was supplemented by naval gunfire from one cruiser, two gunboats, four monitors, and two torpedo-boat destroyers.

Zero day was fixed for October 31st, by which time the Desert Mounted Corps and the XX Corps were to be as near as possible to Beersheba, while everything was done to contain the enemy in Gaza and to make them think that the main attack was to be against their right flank. Vacated camps were left standing, the navy was active with soundings near the coast, while wells were developed in the Wadi Ghazze, and railway and pipe-line were pushed out in an easterly direction, and the Desert Mounted Corps and XXth Corps gradually advanced in a south-easterly direction down the Wadi Ghazze towards Beersheba.

During this period the Turks made a reconnaissance in force. On October 27th two regiments of cavalry and 3,000 Turkish infantry with guns attacked two advanced posts of our outposts, held by the London Yeomanry of the 8th Mounted Brigade on Hills 630 and 720 near Girheir. The post on Hill 720 was overwhelmed, but the post on Hill 630 held out till relieved by the advance of the 3rd A.L.H. Brigade and the 158th Brigade.

The Turks appear to have gained no information from this operation. Our preparations continued without further interruption. By the night of October 30th/31st our troops were located as follows: The Anzac Mounted Division was at Asluj; the Australian Mounted Division was at Khalasa; the 7th Mounted Brigade was at Esani. The Yeomanry Division, acting as a covering force in the centre, was at Abassan el Kebir. The XXI Corps was opposite Gaza. The XX Corps, forming part of the striking force, was disposed as follows: The 10th Division and Camel Corps Brigade were at Shellal, the 53rd Division was at Goz el Goleib, the 60th Division was at Esani, the 74th Division was at Khasif. The details of the attack on Beersheba were for the 74th and 60th Divisions to attack Hill 1070 and the enemy's works between the Wadi Saba and the Khalasa Road on a front of 5,000 yards. Part of the 53rd Division covered the left of this attack. The 7th Mounted Brigade covered the right of the 60th Division.

The infantry were to be in a position about 2,500 yards from the Turks' trenches, from which they could assault their works by 0400 hours on October 31st. The mounted divisions were to be east of Beersheba early enough to attack it before the enemy realized the attack of the XX Corps. The main attack of this corps, it was anticipated, would be delivered between 1000 and 1100 hours on zero day. The G.O.C. Desert Mounted Corps was told to keep his men and horses as fresh as possible for the principal operations, in which they were to pass round the Turks' left flank and gain a position in the vicinity of the Tel Nejile—Wadi Hesi.

On October 31st the Turks were surprised by the direction of the attack of our mounted troops from the east of Beersheba after their night marches of twenty-five and thirty miles respectively. They reached their first objectives east of Beersheba, in the vicinity of Khashm Zanna by 0800 hours. By 0830 hours Hill 1070 had been captured by the 181st Brigade. Our guns were then moved forward to wire-cutting range of the enemy's main position between the Khalasa Road and the Wadi Saba. This position was bombarded from 1030 hours until midday, and then the 74th Division successfully assaulted it. By 1930 hours the Turkish defences north of the Khalasa Road were captured by the reserve brigade of the 74th Division. During these operations by the infantry, 500 Turks and six field guns had been captured.

On arrival at their first objective the mounted divisions had to cross an open plain commanded on the north-east and south-east and flanked by Tel Saba and Tel es Sakaty. It was not till 1300 hours that Sakaty was captured by the 2nd A.L.H. Brigade. Saba did not fall until 1500 hours. The 4th A.L.H. Brigade was ordered to make a mounted attack against the trenches covering Beersheba. They were supported by "A" Battery H.A.C., and the Notts Battery, firing at a range of 2,500 yards. By 1830 hours the 4th A.L.H. Brigade had captured the Turks' trenches. Their forward squadrons galloped over the two front lines of trenches, then dismounted and attacked the occupants with the bayonet; the remainder of the Brigade galloped into the town and captured 1,100 prisoners and ten guns of the Turkish 27th Division. Also they prevented the Turks from destroying more than two of the seventeen wells in the town. The 4th A.L.H. Brigade lost 31 killed and 33 wounded.

This preliminary operation was thus completely successful owing to the fact that the Turks were surprised, and that the final assault was carried through with great determination and rapidity.

Now that the Turkish left flank at Hureira and Sheria was exposed, and the XX Corps was within striking distance of it, it was essential to deliver the main attack as early as possible.

It was also necessary to contain the Turkish 3rd and 53rd Divisions in Gaza to draw their reserves in this direction, and also to hold off the troops north of Beersheba while the XX Corps had time to reconnoitre the enemy's main position and to assemble for the attack. Accordingly, the XXI Corps was to capture the line Umbrella Hill-Shaikh Hasan, on a front of 6,000 yards to a depth of 3,000 yards. The 53rd Division, Camel Corps Brigade, and Anzac Mounted Division occupied a line Bir Marrineh-Abu Jowal on November 1st. At 2300 hours on this day the 156th Brigade captured Umbrella Hill. By 0300 hours on November 2nd the 161st and 162nd Brigades attacked on a front of 6,000 yards, and by 0630 hours had reached Shaikh Hasan.

The Turks had lost heavily during our preliminary bombardment of Gaza and its vicinity since October 26th, and in consequence these attacks on November 1st and 2nd were carried out with little loss. During these days, however, the Turks were preparing for a counter-stroke north of Beersheba. On November 3rd the 53rd Division moved towards Khuweilfeh Hill. Here strong opposition was encountered from three cavalry regiments and eight battalions. The Turks continued their attacks throughout November 4th and 5th in their attempt to drive back our covering force on Beersheba, and to induce the Commander-in-Chief to alter his plans and to make his main attack against them in the Hebron Hills. The Commander-in-Chief, however, continued with the plans and preparations for his main objective, which was to attack the enemy in the Sheria-Hureira position on November 6th. In this connection the Commander-in-Chief writes in his despatches: " Had the enemy succeeded in drawing considerable forces against him in that area the result might easily have been an indecisive fight, and my own striking force would probably have been made too weak effectively to break the enemy's centre in the neighbourhood of Sheria and Hureira. However, the enemy's action was not allowed to make any essential modification to the original plan." General Barrow was given command of the right flank guard during the main attack. His force consisted of the 53rd Division with the Camel Corps Brigade, the Yeomanry Division, the New Zealand and 2nd Australian Light Horse Brigades. This force maintained its position, and when the 60th Division successfully attacked the Turks' entrenched position at Sheria the enemy had no reserves left to re-establish the situation, and, in consequence on November 7th their whole defence collapsed. By nightfall on this day only their troops in the Atawineh works still held out. It had, however, been hoped that Sheria and the water in the Wadi would have been captured on November 6th. But the 60th Division was unable to do this. One brigade of the 10th Division was able to advance to within a mile of Hureira. The remainder of the division remained east of the Sheria-Beersheba railway. The 74th Division was a mile north-east of the 10th Division by the evening of November 6th. On that night the 75th Division captured Outpost and Middlesex Hills at 2330 hours and Turtle Hill at 0500 hours on the next day. Only the troops in the Atawineh works still held out. Also, early on November 7th, the 10th Division captured Hureira, and the 60th Division, after capturing Tel el Sheria, advanced two miles beyond the Wadi Sharia. The cavalry now passed through the infantry to join up with the XXI Corps and to prevent the Turks at Atawineh from escaping north. Owing, however, to the difficulties of water supply the mounted troops became widely distributed when the opportunity for pursuit arrived and so the bulk of the Turkish 26th and 54th Divisions gained a position north of the Wadi Hesi before the Desert Mounted Corps was able to join up with the XXI Corps.

The Turkish retreat, however, was energetically followed up by the 60th Division as far as Huj. The 54th Division advanced through Gaza, and north-west of it through Sheikh Redwan to the sea, the Imperial Service Cavalry Brigade advanced up to Beit Hanun, and the 52nd Division advanced up the sea coast past the 54th Division to the Wadi Hesi.

The Turks had thus been successfully driven from their naturally strong position which they had been fortifying since the first battle of Gaza.

The Commander-in-Chief gives full credit to the preparations which had made this feat possible. In his despatch of June 28th, 1919, he wrote: " I desire to express my indebtedness to my predecessor, who, by his bridging of the desert between Egypt and Palestine, laid the foundations for the subsequent advances of the Egyptian Expeditionary Force. The organization he created, both in Sinai and in Egypt, stood all tests and formed the corner-stone of my successes."

The orders issued by the Commander-in-Chief now were for the Desert Mounted Corps to press on by the shortest route with all available forces to Tireh-Beit Duras, supported by the XXI Corps advancing towards Julis-Mejdel, as quickly as possible. The main objective given was the junction of the railways to Beersheba and Huj from Jerusalem and

Tulkeram.   The supply question limited the number of divisions which could follow up the enemy; all the available transport had to be transferred to the 52nd and 75th Divisions, who now took up the pursuit with the Desert Mounted Corps.   The Turks on November 10th were on a front of twenty miles between Jibrin and Kubeibe, covering Junction Station with small parties detached to Hebron and to Dhahariye.   A pause had to be made to organize the pursuing force for attack.

The 10th, 74th, and 60th Divisions were sent back to Belah to rest, and to be refitted.   The 54th Division was deprived of all its transport and left at Gaza.   The 53rd Division concentrated about Beersheba.   On November 12th the whole Desert Mounted Corps was in action for the capture of Berkusie, Summeil, and Zeita.   The Turks strongly counter-attacked on the right flank, and forced the Australian Mounted Division back to Menshiye.

On the following day the battle for Junction Station was fought.   The key to the position was the villages of Katrah and Mughar, separated by the Wadi Jamus, standing out from the low ground separating them from the rising ground to the west of which stands the village of Beshshit.   At these two naturally strong villages the Turks made most resistance against our turning movement directed against their right flank.   The natural strength of the Turks' position was some compensation for its tactical insecurity owing to the extension of their front, and owing to the fact that it was parallel to the main line of communication running north to Tulkeram.   General Allenby's orders for the operations on November 13th were for the Australian Mounted Division to attack the southern part of the enemy's position to the south of the Gaza-Junction Station road; the 52nd and 75th Divisions were to attack the centre of their position between the Gaza-Junction Station road and Katrah, and seize the Jerusalem railway north of the road.   The Yeomanry Division supported by the Anzac Division was to operate on the left flank of the infantry, and the Camel Corps Brigade was to work along the sandhills in the coastal area.   When Junction Station was captured, the mounted troops were to advance in a northerly direction towards Ramleh, Ludd, and Jaffa.   At first the advance was carried out successfully and with little opposition.   Beshshit was taken by the 52nd Division. The Yeomanry Division occupied Yebna.   The 75th Division captured Mesmiyeh after considerable fighting.   By 1300 hours, however, the 52nd Division was held up in front of Katrah and Mughar, and the Yeomanry Division was similarly checked at Zernukah and Kubeibe.   Orders were sent to the 6th Mounted Brigade, co-operating on the left of the 52nd Division, to attack Mughar from the north.

The Brigade Commander accordingly reconnoitred the ground in his front and assembled his brigade in the Wadi Jamus, while the Berks Horse Battery was located in the vicinity of Beshshit, 3,200 yards from the enemy's position. The machine-gun squadron was posted in the Wadi Ghor close to the 155th Brigade. At 1500 hours the Bucks and Dorset Yeomanry emerged from the Wadi Jamus and advanced in column of squadrons towards the hill just north of the village of Mughar. The troops were exposed to view and fire the whole of the 3,000 yards they had to cross before reaching their objective. In spite of that both regiments gained the crest of the ridge, which was their first objective. Then, with the help of two squadrons of the Berkshire Regiment and the 155th Brigade making a simultaneous attack from the Wadi Ghor, the position was gained by 1700 hours at a loss to the enemy of 1,096 prisoners, two guns and twelve machine guns. The 22nd Mounted Brigade exploited the success as far as Akir, and the 8th Mounted Brigade captured Yebna.

On the following day, at 0730 hours, Junction Station was captured by the 237th Brigade of the 75th Division and the 12th Light Armoured Motor Battery. The Turkish Armies were now forced apart down their two lines of communication.

Their Seventh Army withdrew towards Jerusalem, and their Eighth Army retired in a northerly direction. No good communication between the separated forces of the enemy existed south of the line Tulkeram-Nablus.

The following is a summary of the present situation from despatches: " In fifteen days our force had advanced sixty miles on its right and 40 miles on its left. It had driven a Turkish army of nine divisions out of a position in which it had been entrenched for six months, and had pursued it, giving battle whenever it attempted to stand, and inflicting on it losses amounting probably to nearly two-thirds of the enemy's original effectives." The Commander-in-Chief, after the occupation of Jaffa, decided to advance on Jerusalem at once against the Turkish Seventh Army with the forces which he could keep supplied from those available locally. It was a difficult decision, as some troops would have to be sent north to contain the Eighth Turkish Army in the Plain of Sharon. His troops were tired and had had severe casualties; transport resources were limited, as borrowed transport from the 54th Division had now to be returned; there was only one road fit for wheeled transport, and for nearly four miles this road between Bab el Wad and Saris passed through a narrow defile damaged in many places by the Turks. The Turks also had received reinforcements.

The plan now was to send two infantry and one mounted division into the hills towards Jerusalem, and to cover the lines of communication in the plain with one infantry and one mounted division. On November 16th Jaffa was occupied by the Australian and New Zealand Mounted Division, while the Australian Mounted Division advanced on Latrun. Orders for the advance on Jerusalem were for the 75th Division to advance up the Latrun-Jerusalem road with the 52nd Division on its northern flank, marching in tracks three miles north of the main road, and the Yeomanry Division covering its left flank. It was hoped to isolate Jerusalem by getting astride the Jerusalem-Nablus road north of the city, and thus to avoid fighting in the vicinity of sacred ground. By November 21st the weather had broken. There had been considerable difficulty in getting the guns forward. The yeomanry were in very rough and broken ground and had suffered many casualties. Their horses were an encumbrance, and, in spite of determined efforts on foot, they were unable to capture the Zeitoun Ridge, on the eastern end of which lay the village of Beitunie, strongly held by some 3,000 Turks supported by field guns. The yeomanry had only four mountain guns for their close support. The 52nd Division was unable to capture El Jib. Nebi Samwil, however, was taken by the 234th Brigade. This was the limit of our advance towards Jerusalem. We had, however, secured the difficult passes through the hills. The Commander-in-Chief writes in despatches as follows: " The troops had achieved valuable results. Had the attempt to force the narrow passes from the plain to the plateau of the Judaean range not been made at once, or had it been pressed with less determination, the enemy could have had time to reorganize his defences in the passes low down, and the conquest of the plateau would have been slow, costly, and precarious. As it was, positions had been won from which the final attack could be prepared and delivered with good prospects of success."

In the coastal area the situation was satisfactory. On November 24th the Australian and New Zealand Division secured a bridgehead over the River Auja with two companies of the 54th Division in order to contain the Eighth Turkish Army and to prevent them from reinforcing their troops in front of Jerusalem. A position was gained north of Hadrah and Muannis on November 27th. At dawn on the following day the 3rd Turkish Division regained their positions on the north bank of the River Auja. It was then evident that the establishment of large posts on the right bank of the river would be necessary if bridgeheads were to be maintained there. Our offensive operations towards Jerusalem were not

continued for a fortnight. During that interval the XX Corps
was brought up, and the XXI Corps in the hills was relieved
and sent to the plains between Ramleh and Jaffa. The
yeomanry were relieved by the 74th Division, and went into
reserve south of Jaffa. By December 6th the 53rd Division
was four miles south of Bethlehem, in touch with the
10th Australian Light Horse Brigade in the Wadi Sura. The
60th Division was south of the Jaffa-Jerusalem road, the
74th Division was in the vicinity of Nebi Samwil, the 10th
Division was between Sira and Beitunie, facing north-east,
and the Australian Mounted Division was between the XX
and XXI Corps. On December 8th the main assault started
at dawn. Troops quickly gained their first objective. Owing,
however, to the difficulty of obtaining close artillery support
in this hilly country the further advance was slow. By midday, however, the 74th Division had captured Beit Iksa, and
the 60th Division had gained a commanding position east of
Wadi Sura. The 53rd Division had been delayed by the difficulty of the ground, and thus the left flank of the 60th Division
was exposed on reaching a position one and a half miles west
of Jerusalem. When the attacks ceased on this day, all the
dominating positions west and north-west of the enemy's main
line of retreat to the north had been captured. The Turks
were now fighting on a line parallel to their line of communication between Jerusalem and Nablus. On December 9th the
Turks made no effort to recover their lost positions. Early
in the morning the Worcester Yeomanry gained a position
astride the Jerusalem-Jericho road. The 53rd Division
occupied the Mount of Olives and the high ground east of it.
The 60th and 74th Divisions advanced to their final objectives,
and thus a complete cordon was drawn round Jerusalem.
The Mayor of Jerusalem, with a flag of truce, arranged for
the surrender of the city early on this day. On December 11th
the Commander-in-Chief made his official entry into Jerusalem.
The moral significance of this success was considerable. In
addition, the last Turkish reserves had been used up and
danger to Baghdad had been removed. The Arabs became
more loyal and energetic in the prosecution of their harassing
methods against the Turkish communications. The point for
consideration now was whether any further advance on a big
scale should be carried out on this front once the line covering
our positions from Jerusalem to the River Auja had been
consolidated. The Turks had fewer men than we had, and
even an advance to Aleppo would not help the situation on
the Western Front, where it was vital for us to have every
available man in order to win the war. A victory over the
Turks would not help us to gain our main objective. The

campaign, if continued offensively, would be costly, and a great drain on our man-power and resources. On the other hand, it was considered that, if Turkey was defeated and forced out of the war, Bulgaria would follow, and the Germans might realize the failure of their eastern policy, and if they also failed to break through our front on the West then they might make peace. It was therefore decided that our offensive policy against Turkey was to be continued. For the consolidation of our position it was first necessary to render Jaffa secure for shipping. The XXI Corps was directed to drive the Turks from the north bank of the River Auja and to occupy the line Rentia-El Jelil. The crossing of the river was made by the 52nd Division on the night of December 20th/21st. The Turks were completely surprised when the crossing of the covering party was effected by 2230 hours on December 20th. By 2300 hours the first bridge was across the river. By midnight the whole of the 156th Brigade was across. By the morning of the 21st all three brigades were across and on the Turkish line dominating the river. At a cost of a hundred casualties we captured 316 prisoners and ten machine guns. On the right of the 52nd Division the 54th Division captured Bald Hill, and on the 22nd exploited this success up to Mulebbis and Rentia. The 52nd Division, in order to secure Jaffa, pushed the enemy back eight miles from this port; their left was up to Arsuf. Our advance in front of Jerusalem was postponed when information was received that the Turks were planning a counter-attack for the recapture of this city. The Turkish attack began on the night of December 26th/27th, and was continued until the afternoon of the 27th, when we advanced 4,000 yards on a six-mile front to Sheikh Abdullah, Jeirut, the Zeitoun Ridge, and Dreihemen. The method of dealing with the Turks' attack was to hold them on the front of the 53rd and 60th Divisions, and to attack with the 74th and 10th Divisions. The Turks made their most determined efforts for the capture of Ful Hill, held by the 53rd Division throughout the day. On December 28th, having made their effort and lost heavily in their failure, the Turkish rearguards were driven back by the XX Corps.

Hostile resistance weakened as our troops reached their final objectives on December 30th. In these last three days' fighting the Turks' casualties were more than 4,000; 750 prisoners and twenty-four machine guns had been captured, and an advance of six miles in the vicinity of Bireh and three miles from Tahta had been made. It was necessary now to improve roads and to bring the railways nearer to our front. The XX and XXI Corps now satisfactorily covered Jerusalem, Ramleh, and Jaffa, and the roads joining these places. During the past six weeks the Egyptian Expeditionary Force had advanced its front approximately sixty miles. There were no further active operations on a large scale until February, 1918.

1918. The Cabinet's policy was still to continue offensive operations against the Turkish Army. The Commander-in-Chief, however, made it clear that the supply factor had to be considered, and that the Egyptian Expeditionary Force had outrun its communications, so that an immediate resumption of any forward movement was not for the present possible. The railway was fifteen miles south of Ludd on January 1st. The Turks had effectively destroyed the big bridges on the Jaffa-Jerusalem railway, and there was great difficulty in feeding the troops and the large population in Jerusalem by a single road over difficult passes. Also, before an advance could be made to any distant objective, reinforcements might be required, as the Turks would be nearer to their sources of supply and to their broad-gauge railhead at Rayak. Many preliminary operations were necessary before an offensive on a large scale could be undertaken.

The Commander-in-Chief wished to make an advance to secure our right flank by driving the Turks beyond the River Jordan, and to obtain a starting point for raids into Moab and for an attack on the Hedjaz line in co-operation with the Arabs based on Aqaba. By securing the crossings over the Lower Jordan it would be possible to gain control of the Dead Sea and the fertile districts about Kerak, and to prevent the Turks from making raids on our positions west of this sea. The operations began on February 19th with an advance to the River Jordan and to the Wadi Auja, eight miles north of Jericho. The troops employed were the 60th, 53rd, and the Australian and New Zealand Divisions. The 60th Division was to advance on Jericho, the Australian and New Zealand Division was to turn the Turks' left flank from the vicinity of Nebi Musa. Our progress was slow and difficult. The Turks were not driven across the River Jordan till the night of February 20th/21st, leaving a bridgehead on the west bank of the river at Ghoraniyeh. On February 21st the 1st Australian Light Horse Brigade entered Jericho, and the New Zealand Brigade occupied the Turkish base on the Dead Sea. Any threat to Jerusalem from the east was now effectually removed. But it was still necessary to drive the enemy farther northwards on both sides of the Jerusalem-Nablus road in order to control the routes into the Jordan Valley, and to add to the difficulties of the Turks in dealing with any force advancing towards Amman to co-operate with Feisal's Army.

Our further advance began on March 8th. The objective for the XX Corps was Abu Tellul, north of the Wadi Auja, Kefr Malik, Sinjil, north-west of the Wadi Jib and Nebi Saleh. The 54th and 75th Divisions were to capture Wadi Ballut and Ras Ain. When all these objectives had been gained by the two corps after hard fighting by March 12th we had gained strong positions from which operations could be undertaken east of the River Jordan in conjunction with the Arab Army. The Commander-in-Chief now decided to make a raid on the Hedjaz railway, and to destroy the viaduct and tunnel at Amman thirty miles north-east of Jericho. The operations carried out by Feisal's Arab Army against the Turkish communications on the Hedjaz railway had been of great value. They had deprived the Turks of supplies and had caused them to form detachments. Also, the Commander-in-Chief wished to draw as many of the Turks to the east of the River Jordan, and induce a belief in their commander's mind that he intended to operate with the Arabs up the Hedjaz railway. This would facilitate his task when he made his next offensive operation in country best suited for the mounted troops, in which he had a great superiority. His intention was to break through the enemy's front on the coast, where his cavalry could advance up the coastal plain. For these reasons he carried out two raids, which had the effect of causing the Turks to be alarmed about their left flank, and to leave a third of their forces east of the River Jordan. They did not withdraw troops to reinforce their troops on the right of their line. The raid to Amman started on March 21st and lasted until April 2nd. The troops allotted to the raid were under the G.O.C. 60th Division, and consisted of the Australian and New Zealand Division, the Imperial Camel Corps Brigade, a mountain artillery brigade, a light armoured-car motor brigade, a heavy battery, and the 60th Division. The plan was for the 60th Division to force the crossings at Hajlah and Ghoraniyeh over the River Jordan with a mounted brigade co-operating on its left flank. An advance was then to be made up the road to Salt after Shunet Nimrin had been captured.

The rest of the force was to move direct on Amman. Heavy rains added to the difficulties of crossing the River Jordan, which ran at eight miles an hour. The tracks were rough and unsuitable for wheels, and the troops had to march in single file. It was not until the evening of March 28th that Salt was reached by the 1st A.L.H. Brigade.

On March 26th the cavalry captured 170 prisoners near Sweileh.

A demolition party blew up the railway seven miles south of Amman. On the following day the 181st Brigade, less two battalions, reached a position two miles east of Sweileh. Demolition parties destroyed the railway at Libban and seven miles north of Amman. On March 28th two battalions of the 181st Brigade, the N.Z. Mounted Brigade, supported by the 9th Mountain Brigade, attacked Amman. This attack did not progress owing to the strength of the enemy's position at Amman and at Hill 3039, as the enemy was not surprised, and owing to the difficulty of artillery support and observation.

The enemy, who were 4,000 strong, supported by fifteen guns, were able to hold their positions while their reinforcements tried to envelop the left flank of the 181st Brigade and also attacked our 179th Brigade at Salt. The delay in crossing the River Jordan had enabled the Turks to get up reinforcements for Amman, which now contained approximately 5,000 rifles and fifteen guns. For the next four days, under most adverse conditions and with the artillery support of only three mountain batteries, efforts were made to capture Amman and the positions covering the tunnel and viaduct. But the positions were too strong and were too tenaciously held. After the railway north and south of Amman had been broken, the withdrawal began on the night March 30th/31st. The results of the raid were that the Turks reinforced Amman, and maintained a larger garrison there than formerly. They became apprehensive for the safety of their left flank and of Deraa, and of their communications east of the River Jordan. Therefore, the raid fulfilled its main purpose. While this raid was taking place the German offensive was being successfully carried out against our Fifth Army in France. All our resources, it was seen, would be required to resist the Germans, and, consequently, the Commander-in-Chief was told that a defensive policy would now be adopted in Palestine, and all the British troops that could be spared must be sent to the Western Front. During April and May two divisions, twenty-four British battalions, nine yeomanry regiments, five and a half siege batteries, and five machine-gun companies were dispatched to France from the E.E.F. The yeomanry were replaced by Indian cavalry from France, and the twenty-four battalions by Indian units from India. The climate suited the Indian troops, and their supplies and reinforcements avoided the submarine zone. The reorganization of the E.E.F. was complete by the end of the summer, when it was decided again to resume the offensive.

In the meantime the Turks had retaliated after our Amman raid with an attack on April 11th on our Ghoraniyeh bridgehead and on our positions on the Jericho-Beisan road. They were unsuccessful in these attacks and had heavy casualties. The attackers withdrew to a position at Shunet Nimrin (nine miles south-west of Salt). The Commander-in-Chief determined to make another raid east of the River Jordan in order to facilitate his main operations later in the coastal area when the time came to carry out the offensive on a large scale.

He wished the Turks to continue to keep a large part of their force east of the River Jordan, and to strengthen their belief that he meant to operate up the Hedjaz railway. His plan was to co-operate with the Beni Sakhr Arab tribe to capture Salt with his mounted troops, and to cut the communications of the Turks' isolated force at Shunet Nimrin. Shunet Nimrin would then be attacked and captured by infantry. It was hoped then to hold Salt with the Arabs, and avoid the great summer heat and discomforts of the Jordan Valley.

The attacking force, consisting of the A. and N.Z., Australian, and 60th Divisions and the Imperial Service Cavalry and Infantry Brigades assembled on the Jordan on the night April 29th/30th. On April 30th the 60th Division, with their left flank protected by one brigade A.L.H., attacked Shunet Nimrin, while the 3rd A.L.H. Brigade advanced via Jisr Damie on Salt, which they captured. The 4th A.L.H. Brigade took up a position facing west astride the road to Salt from Makhruk over the Jisr Damie, with patrols up to the Wadi Zerka, and a detachment at Es Shert. The 5th, 1st, and 2nd A.L.H. Brigades during the night April 30th/May 1st closed up on to Salt. On May 1st a Turkish cavalry division and part of their 24th Division, having crossed the River Jordan at Jisr Damie, attacked the 3rd A.L.H. Brigade, which had to fall back to Wadi Abyad with a loss of nine guns. The 60th Division was unable to capture Shunet Nimrin. The co-operation of the Beni Sakhr Arab tribe did not materialize. They failed to keep their promise of holding off Turkish reinforcements coming from Amman to Shunet Nimrin via Ain Sir. On the following days the 60th Division was unable to capture Shunet Nimrin, and the position of the isolated troops at Salt was dangerous, as reinforcements could not get through to them in face of the opposition encountered in the hilly country between Shunet Nimrin and Salt.

The evacuation of Salt was accordingly ordered on the afternoon of May 3rd.

By the evening of May 4th the raiding force was west of the River Jordan, having crossed at Ghoraniyeh. Though our losses were approximately 1,500, the strategical effect had been favourable, as the Turks kept their Fourth Army, i.e., a third of their force, east of the River Jordan for the rest of the campaign.

The Desert Mounted Corps now had to remain during the summer in the oppressively hot Jordan Valley, 1,200 feet below the level of the sea. The retention of this corps in the Jordan Valley throughout the summer confirmed the Turks in their idea that future offensive operations would be against their left flank. The further operations before our final offensive took place were as follows:—On June 8th, Arsuf (ten miles north of Jaffa) was captured. By July 6th the Arab Army had isolated Medina by their demolition of the railway north of this place. On July 14th, 3,000 Turks attacked and seized Abu Tellul, nine miles north of Jericho. The 1st A.L.H. Brigade counter-attacked and re-took the hill with 338 Turkish prisoners and eight machine guns. By August the strength of the Arab Army was approximately 8,000, with 52 guns, 150 machine guns, and 2,000 riding camels, with which they formed a mounted camel column, including gunners, engineers, infantry, and machine gunners. These forces were a valuable asset on our right flank during our final advance to Aleppo.

In preparation for the final offensive the communications were improved. A standard gauge railway was constructed from Ludd to Jerusalem, and the double line from Qantara to Rafa was extended through Ludd to Jaffa. A light railway was made from Jaffa to Arsuf. The port of Jaffa was improved. The roads within our area were widened. By the end of August the E.E.F. was ready to carry out the plans for a continuous offensive against the Turks until their field armies were finally defeated. The Commander-in-Chief's plan was to concentrate five infantry and three cavalry divisions unknown to the enemy on the coast opposite the weak Turkish right flank. Then, when the infantry had broken through the enemy's lines in the coastal area, and having secured the crossings over the Nahr Falik and then wheeled to the east, the Australian Mounted, 4th and 5th Cavalry Divisions were to advance along the coast to cross by the passes near Megiddo, and to enter the Plain of Esdraelon at Lejjun and Shusheh, and to secure Afule and Beisan and also to seize Jisr Mejamie Bridge over the River Jordan and to send a detachment to Nazareth. The three Turkish Armies on our front totalled 4,000 sabres, 32,000 rifles, and 400 guns. Our XX and XXI Corps with the French contingent comprised 12,000 sabres, 57,000 rifles, and 540 guns. The reason for the area of operations being west of the River Jordan is summed up as follows in the Commander-in-Chief's despatches of October 31st, 1918. "I was anxious to gain touch with the Arab forces east of the Dead Sea, but the experience gained in the raids against Amman and Salt, in March and May, had proved that the communications of a force in the hills of Moab were liable to interruption as long as the enemy was able to transfer troops from the west to the east bank of the Jordan.

D

This he was in a position to do, as he controlled the crossing at Jisr Damie. The defeat of the Seventh and Eighth Turkish Armies, west of the Jordan, would enable us to control the crossing. Moreover, the destruction of these armies, which appeared to be within the bounds of possibility, would leave the Fourth Army isolated if he continued to occupy the country south and west of Amman. I determined, therefore, to strike my blow west of the Jordan." Once the important positions on the Turks' lines of communication at Afule and Beisan were in our possession we should block the line by which the Turkish Seventh and Eighth Armies could retreat. These places could be more quickly reached by the Plain of Sharon than by the mountainous country north of Jerusalem. The advance by the coastal route would enable the cavalry to cross the mountains of Samaria at their most narrow point; also an advancing army could be more easily maintained in this area than in the hills. The British numbers for the main attack on our left flank were 9,000 sabres, 383 guns, and 35,000 rifles of the XXI Corps, 60th Division, and Desert Mounted Corps (less one division) to break through the Turkish defences on a frontage of 4,000 yards to a depth of 3,000 yards between Bir Adas and the sea, as well as their second line nearly 3,000 yards farther north between Tireh and the south of the Nahr Falik.

The Turks held these positions with 1,200 sabres, 8,000 rifles, and 130 guns. Our concentration was complete by the evening of September 18th.

The attack was launched at 0430 hours on September 19th. The Turks were completely surprised. The steps taken to gain this surprise were as follows: R.A.F. prevented the enemy from flying over our positions by day. Camps and dummy grazing guards were arranged behind the lines of the XX Corps. These guards were active during the day. Headquarter Camp of the Desert Mounted Corps remained in its position in Talaat Dumm long after the Commander and Staff had moved to the coast. An advanced report centre for G.H.Q. was constructed at Jerusalem, as if to be in touch with operations on the eastern flank. Normal routine was maintained in vacated camps by personnel of the British West Indian battalions observed by the Turks from Shunet Nimrin. The Arab flying column on September 16th attacked the Hedjaz railway sixteen miles south of Deraa. On September 18th the Flying Corps bombed Deraa station. The result was that the attack of the XXI Corps, after the intense bombardment lasting till 0445 hours, was completely successful and everything was carried out according to plan. The French contingent, 3rd, 7th, 54th, 60th and 75th

Divisions, quickly overcame the enemy's resistance at Wadi Karah, Bir Adas, Tabsor, and along the coast. Their second line of defence was then overwhelmed, and when the 60th Division had cleared the Nahr Falik and the marshes in the vicinity of the enemy, and advanced north-east on Tulkeram, the mounted troops at 0630 hours began to pursue the retreating Eighth Turkish Army.   The orders to the mounted troops were to disregard hostile troops not directly attempting to check their advance.   Thus, twenty-four hours after the attack had started the lines of retreat of the Turkish Seventh and Eighth Armies were blocked.   The offensive was continued towards Nablus by the 53rd and 10th Divisions against the Seventh Army as soon as the Eighth Army had been driven back.   By midday on September 20th the 10th Division had advanced seven miles.   Messudie, which was on the Turks' line of retreat towards Jenin, was secured. On the following day Nablus was secured, and the Seventh Turkish Army was in full retreat.   The offensive was continued so vigorously by the infantry that all organized resistance by the enemy rearguards ceased on September 21st.   The cavalry, who were astride their main line of retreat by September 23rd at Afule and Beisan, with Haifa, Acre, Tiberias, and Jisr Mejamie Bridge in their possession, were able to deal effectively with the retreating Seventh and Eighth Armies, which, as armies, ceased to exist by September 24th.   The Desert Mounted Corps was now ordered to cut off the retreat of those of the Fourth Army who had escaped from Chaytor's force and to occupy Damascus.   On September 25th the A. and N.Z. Division captured Amman.   The advance was now made in two columns.   The Australian and 5th Cavalry Divisions from Nazareth marched via Tiberias, and crossed the River Jordan, south of Lake Huleh, after stiff fighting at Benat Yakub Bridge.   They reached Kuneitra on the 28th, Sasa on the 29th, and the vicinity of Katana on the 30th.   The 4th Cavalry Division was at Irbid and Remte on September 27th, on the 28th at Mezerib, on the 29th in touch with the Arab mobile force at Ezra.   At dawn on October 1st the 1st and 3rd A.L.H. Brigades entered Damascus and began to advance on Homs.   In this city 12,000 Turks were collected.   The hostile forces, now reduced to approximately 17,000, were a rabble.   They had no artillery, organization, or transport.   The Commander-in-Chief determined to exploit his success.   His next bound was to be to the line Rayak-Beirut.   The 7th Division marched to Haifa, which was reached on October 1st.   Their march was continued to Beirut, which was reached on October 8th.   The 5th Cavalry Division with an Arab force now continued the advance from Rayak on Aleppo.   The 5th Cavalry Division

D 2

was now divided into two columns operating a day's march apart. The leading Column "A" consisted of three armoured-car batteries, 3 light-car patrols, and two regiments of the 15th Brigade. Column "A" reached Homs on October 16th. On the 18th a brigade of the 7th Division reached Tripoli. The occupation of this place, where stores could be landed and taken by motor lorry to Homs, eased the supply situation for the mounted troops north of the Lebanon Mountains. The roads and railway behind the troops were in such a bad state of repair that it was only by occupying Beirut and Tripoli that they could be kept supplied. The Hedjaz railway had been considerably damaged, and there was very little rolling stock for the railway north of Rayak. Although Aleppo was 120 miles from Homs, and 20,000 Turks were reported at Aleppo, and our 5th Division was out of touch with the rest of the Desert Corps, it was decided to continue the advance. On October 21st Column "A" reached Zor Defai. On the following day the enemy rearguards were driven from positions near Khan Sabil, half-way to Aleppo. On the next day the armoured cars were within a few miles of Aleppo. The Turks were summoned to surrender the town. This summons was refused. Before making an attack it was, therefore, necessary to wait for the remainder of the 5th Cavalry Division to come up and also to reconnoitre their positions. However, an attack was not necessary, as the Arabs were able to enter the town on the evening of October 25th. Our armoured cars followed at 1000 hours on the following morning. Our final operation was a charge made by the Jodhpur and Mysore Lancers against 3,000 Turks, who, however, counter-attacked and forced our cavalry to occupy a defensive position. Later in the day the Turks continued their retreat. The situation is summed up in Despatches as follows: " Between September 19th and October 26th, 75,000 prisoners were captured. In addition, 360 guns fell into our hands, and the transport and equipment of three Turkish armies. The 5th Cavalry Division covered 500 miles, and captured over 11,000 prisoners and fifty-two guns." During the thirty-eight days of operations our battle casualties had been less than 5,000.

The Armistice between the Allies and Turkey was completed on October 31st.

THE PALESTINE CAMPAIGN

CHAPTER I.

DIARY OF EVENTS.

Turkey mobilized in August. Fifteen divisions were formed. Later, when war was declared, fifteen more were raised.

1914.

October 30th.—Turkey declared war on Great Britain. They now had four army corps in Armenia for the invasion of the Caucasus. Djemal Pasha had about 140,000 men between Aleppo and Damascus for operations in Mesopotamia or in Egypt.

This added to our difficulties in Persia, Egypt, and India, and to Russia's in the Caucasus. The Black Sea was closed as a means of communication with Roumania and Russia.

Turkey hoped for territorial gains from Russia and Persia.

December 23rd.—The Turkish force organized for operations against Egypt entrained at Damascus. This force consisted of two cavalry regiments, two 6-inch howitzers, two engineer companies, three infantry divisions, and two machine-gun companies.

December 31st.—The 10th and 11th Indian Divisions, 42nd Division and Australian and New Zealand troops had arrived in Egypt.

1915.

January 13th.—Turkish forces arrived at Auja and Kossaima.

February 2nd.—12,000 Turks with guns were distributed between Serapeum, Ferry Post and Qantara.

February 3rd.—Turks were driven back by our fire action supported by five cruisers, torpedo-boats and armed launches in the Suez Canal.

February 5th.—Turks withdrew from the Suez Canal.

April 28th.—Egyptian Expeditionary Force landed in Gallipoli.

1916.

April 23rd.—3,650 Turks, six guns, four machine guns, and six companies of camelry surprised our advanced posts at Oghratina and Qatiya and captured three and a half squadrons of yeomanry and details. Their raid at Dueidar failed.

July 19th.—8,000 Turks from Shellal reached Abd, Bir Jemeil and Bir Bayud.

August 3rd.—Turks occupied Qatiya, Hill 110 and Rabah. During the night the enemy between Katib Gannet and Hod el Enna attacked our outpost line, which was forced to withdraw.

August 4th.—Turks, 18,000 strong with thirty guns, were defeated at Romani. They had 5,000 casualties; 4,000 prisoners were taken and also 4 guns and 9 machine guns.

September 23rd.—Taif was captured by the Arabs. Medina was invested.

December 23rd.—The Turks at Magdhaba were surrounded and 1,280 of their 27th Division with 4 guns were captured by the Australian and New Zealand troops and the Imperial Camel Brigade.

1917.

January 9th.—1,600 prisoners, 4 guns, and 6 machine guns were captured at the frontier post of Rafa by a force of Australian and New Zealand mounted troops and Imperial Camel Corps.

February 5th.—The Senussi were driven out of Siwa.

March 25th.—The Desert Column was concentrated at Belah, the 54th Division was at In Seirat, the 52nd Division at Khan Yunus, and the Camel Corps and armoured batteries near Abassan el Kebir.

March 26th.—First Battle of Gaza.

March 27th.—Our 53rd and 54th Divisions withdrew to the western side of the Wadi Ghazze covering Belah.

April 5th.—The railway reached Belah.

April 17th.—Two divisions crossed the Wadi Ghazze and occupied Shaikh Abbas, Mansura, and Lees Hill.

April 19th.—Second Battle of Gaza. After a loss of 6,400 casualties, our troops consolidated their position from the Wadi Ghazee round Shaikh Abbas, Mansura, Samson Ridge, Shaikh Ajlin. Position warfare then began from Tel Jemmi to the sea.

May 23rd.—A raid was made by the Camel Corps, one brigade, Australian and New Zealand Division, two field squadrons, and a battery of horse artillery, to cut the railway line at Asluj and Hadaj. Thirteen miles of railway were destroyed.

June 28th.—General Allenby became Commander-in-Chief of the Egyptian Expeditionary Force.

The Turkish position now extended to Beersheba. Gaza and Beersheba were both strongly entrenched and wired. In addition, there were redoubts between these places at

Kauwukah, Hureira, Baha, Atawineh, and Sihan on the Gaza-Beersheba road. These redoubts were mutually supporting, and were from 1,500 to 2,000 yards apart, except on the left between Kauwukah and Beersheba, where there was a gap of four and a half miles.

Beersheba, therefore, remained a detached post. Communication trenches, covered by wire, connected up their other posts, except between Ali Muntar and Sihan.

Their troops were the XX and XXII Corps with a cavalry division, viz., one cavalry division of 1,500 sabres, 300 guns, and nine infantry divisions of 50,000 rifles.

The E.E.F. consisted of three mounted divisions, the XX Corps of four divisions (10th, 53rd, 60th, 74th) and the XXI Corps of three divisions (52nd, 54th, 75th) and a composite force.

Our front of twenty-two miles was close to the Turks only near Gaza. It was ten to twelve miles away in other parts.

The reinforcements asked for by General Allenby were 11 batteries, 6 groups of heavy artillery, 2 divisions, and 3 squadrons, R.F.C.

June to October.—Communications were improved. The Egyptian Labour Corps was increased to the number of over 55,000. They worked on road and railway making, loading trains, surf-boats, and ships, and in laying the pipe-line up to Belah. Reservoirs were built at Belah, Jemmi, Sitta, Khan Yunus, Shellal, Rafa, and Qantara.

The Camel Transport Corps of 35,000 camels was organized in companies, each approximately 2,000 strong, to act as first-line transport. Four donkey companies, each of 2,000 donkeys, were formed.

The 140-mile, 4 ft. 8½ in. railway from Qantara to the Wadi Ghazze was extended beyond Belah, and to Karm and Gamli.

Two divisions (10th and 60th) were sent from Salonika, and the 75th Division was formed.

The Turks also improved their communications. The 3 ft. 6 in. railway from Junction was brought to the Wadi Hesi and through Sheria to Beersheba.

October 22nd.—Force Order No. 54 was issued. By this Order the enemy's left flank was to be turned by the Desert Mounted Corps attack with two divisions and the 7th Mounted Brigade from the east and north-east. Two divisions of infantry were to attack the main Beersheba defences from the south-west, while one division covered the left flank of the attacking force. The XXI Corps was left for the attack on Gaza. One mounted division was to hold the centre.

October 26th.—Bombardment of Gaza began. This was continued for eight days.

October 27th.—Two regiments of cavalry and 3,000 Turkish infantry with guns attacked two advanced posts of our outposts held by Middlesex Yeomanry on Hills 630 and 720 near Girheir. The line was held until the 158th Brigade reinforced the yeomanry.

October 28th.—53rd Division and 229th Brigade occupied an outpost line covering the 74th and 60th Divisions at Tel Fara and the construction of the railway extension to Karm. 1st Australian Light Horse Division reached Asluj. The Anzac Mounted Division reached Khalasa.

October 29th.—60th Division and a brigade of the 74th Division came into line with the 53rd Division facing in a south-easterly direction. The 10th Division was west of Shellal. One cruiser, two gunboats, four monitors, and two torpedo-boat destroyers joined in the bombardment on the Gaza defences.

October 30th.—The Australian Light Horse and Anzac Mounted Divisions moved respectively into positions about Khashm Zanna. The 60th and 74th Divisions advanced to a position from which to attack the enemy's positions between the Khalasa road and the Wadi Saba.

October 31st.—The Turks were surprised by the direction of the attack of our mounted troops from the east of Beersheba after marches by the two divisions of twenty-five and thirty miles respectively. They reached their first objectives east of Beersheba by 0800 hours. The 60th Division took Hill 1070 by 0845 hours, and by 1300 hours the objectives for both divisions were captured.

By 1800 hours the mounted troops had captured Beersheba and the greater part of the Turkish 27th Division, and had prevented the Turks from destroying more than two of the seventeen wells in this town.

Colonel Newcombe, with a small mobile force, reached a position between Dhahariye and Hebron in order to cut off Turks retiring from Beersheba.

November 1st.—53rd Division and Camel Corps Brigade and Anzac Mounted Division occupied a line Bir Marrineh-Abu Jowal to protect the northern flank of the rest of the XX Corps in their attack on the enemy's main position between Sheria and Hureira.

The 156th Brigade at 2300 hours captured Umbrella Hill.

November 2nd.—At 0300 hours the 161st and 162nd Brigades attacked on a front of 6,000 yards, and by 0630 hours reached Shaikh Hasan.

The 2nd Australian Light Horse Brigade reached Miljabo and the 7th Mounted Brigade occupied Khuweilfeh.

The 10th Division moved up to Irgeig.

November 3rd.—53rd Division attacked the Turks' position on Khuweilfeh and was able to capture part of it.

During the shortage of water the Australian Mounted Division had to go back to Karm.

November 4th.—Turks with three cavalry regiments and eight battalions heavily counter-attacked the 53rd Division and gained a footing on Khuweilfeh Hill. They were driven off later.

November 5th.—The Yeomanry Division took over the positions occupied by the 74th Division. The Turks continued to attack the 53rd Division without gaining ground.

General Barrow was given command of the right flank-guard during the main attack on the Hureira-Sheria position. This force consisted of the 53rd Division with the Camel Corps Brigade, the Yeomanry Division, the New Zealand and 2nd Australian Light Horse Brigades.

November 6th.—At dawn the main position held by the XX Corps of the Turkish Seventh Army was attacked. The 74th Division on the right was able to capture its objectives. Then the 60th Division pushed on to Sheria Station, but was unable to cross the wadi. One brigade of the 10th Division was able to advance to within a mile of Hureira. The remainder of the division remained east of the Sheria-Beersheba railway. The 74th Division was a mile north-east of the 10th Division, south of the Wadi Sharia and facing north-east.

The 53rd Division drove the enemy off Khuweilfeh.

November 7th.—By 0740 hours the 75th Division had captured Outpost Hill, Middlesex Hill, Turtle Hill, and Ali Muntar. Only troops in the Atawineh works still held out.

The 54th Division advanced through Gaza, and north-west of it through Sheikh Redwan to the sea. The Imperial Service Cavalry Brigade advanced up to Beit Hanun and the 52nd Division advanced up the sea coast past the 54th Division to the Wadi Hesi.

The 10th Division captured Hureira, and the 60th Division, after capturing Tel Sharia, advanced two miles beyond the Wadi Sharia. The cavalry now passed through the infantry to join up with the XXI Corps and to prevent the Turks in the Atawineh position from escaping north.

Owing to enemy opposition and to shortage of water, the cavalry were unable to reach positions farther north than four and a half miles north of Tel Sharia and two and a half miles north-west of it respectively.

November 8th.—The bulk of the Turkish 26th and 54th Divisions gained a position north of the Wadi Hesi before the Desert Mounted Corps was able to join up with the

XXI Corps. The 75th Division captured Beer Trenches, Tank and Atawineh Redoubts, and then joined with the 10th Division.

The Australian and New Zealand Division reached Jemmaneh, where there was a good water supply. Three hundred prisoners were captured here. The Australian Mounted Division, supported by the 60th Division, drove back the Turkish rearguards, and after entering Huj gained contact with the Imperial Service Brigade east of Deir Sineid and south of two brigades of the 52nd Division, prolonging our line north of the Wadi Hesi to the sea.

The last position held by the Turks covering Huj was charged by a yeomanry brigade. The Turks were dispersed and their 12 guns were captured.

The 52nd Division maintained their position north of the Wadi Hesi and drove off four enemy counter-attacks from the direction of Askalon.

November 10th.—Desert Mounted Corps was now ordered by G.H.Q. to " press on by shortest route with all available forces to Tine-Beit Duras supported by XXI Corps advancing towards Julis-Mejdel as quickly as possible."

The main objective given was the junction of the railways to Beersheba and Huj from Jerusalem and Tulkeram.

The Turks were on a front of twenty miles covering Junction Station from Kubeibe-Mughar, five miles west of Junction to Jebrin, with small parties up to Hebron and a detached regiment at Dhahariye. Our main attack was to be carried out against their troops between Kubeibe and Beit Jibrin on November 13th.

The 52nd Division occupied Mejdel. The 157th Brigade reached Esdud after driving Turkish rearguards out of Beit Duras.

Desert Mounted Corps (less New Zealand Mounted Brigade, Yeomanry Division marching north from Huj, and the Imperial Service Cavalry Brigade) covered the front from Faluje to Esdud of the 52nd Division and the 75th Division, now two and a half miles south-east of Mejdel, with a brigade at Suafir. The 54th Division remained at Gaza. The 10th and 60th Divisions withdrew to the vicinity of Karm.

November 11th.—The 1st Australian Light Horse Brigade crossed the Nahr Sukereir and captured a dominating position in Tel Murre. Burka was also captured after heavy fighting. This facilitated the landing of sea-borne supplies.

November 12th.—The whole Desert Mounted Corps was in action on this day in the capture of Berkusie, Summeil, and Zeita. The Turks strongly counter-attacked our right flank,

The Australian Mounted Division was forced to withdraw back on their right flank to Menshiye.

November 13*th*.—The following is a quotation from Despatches: " The country over which the attack took place is open and rolling, dotted with small villages surrounded by mud walls, with plantations of trees outside the walls. The most prominent feature is the line of heights on which the villages of Katrah and Mughar stand out above the low flat ground which separates them from the rising ground to the west, on which stands the village of Beshshit, about 2,000 yards distant. This Katrah-Mughar line forms a very strong position, and it was here that the enemy made his most determined resistance against the turning movement directed against his right flank."

We changed direction to the east in our advance. Yebna was captured by the 8th Mounted Brigade of the Yeomanry Division. Tine was attacked and taken by the Australians. Mughar was captured by the combined action of the 52nd Division and the 6th Mounted Brigade. The infantry had to advance over 4,500 yards of open ground; the cavalry had similarly to cross 3,000 yards. The result of the frontal attack by the infantry and the flank attack by the cavalry was that the position was gained at a loss to the enemy of 1,096 prisoners, 2 guns, and 12 machine guns.

The 75th Division advanced up to Mughar on the railway two miles south-west of Junction Station.

November 14*th*.—At 0730 hours Junction Station was captured by the 234th Brigade of the 75th Division and the No. 12 Light Armoured Motor Battery. The Turkish armies were now forced apart down their lines of communication. The Seventh Army withdrew towards Jerusalem; the Eighth Army retired in a northerly direction.

The first attempt to take Jerusalem had now been definitely held up. Had the enemy, however, not been followed up immediately after the capture of Junction Station they might have been able to organize defences in the narrow passes leading from the plains to the hills. These passes were strong naturally and could have been quickly made formidable. By continuing offensive operations, General Allenby drove the enemy back to the hills until he had gained positions overlooking Jerusalem, and he also secured Jaffa.

There were now no good lines for inter-communication between the two divided Turkish forces covering Jerusalem. The Commander-in-Chief, therefore, decided to advance at once on Jerusalem against the Turkish Seventh Army with the three infantry and two mounted divisions available locally, and to contain the Eighth Army, which had retired north of the River Auja. The difficulties of this situation were that some of our troops would have to be sent north to contain the Turkish Army in the Plain of Sharon; that the Turks had received reinforcements; that the hilly country was unknown and unreconnoitred; that there was only one road fit for wheeled transport leading to Jerusalem; that " for nearly four miles this road between Bab el Wad and Saris passes through a narrow defile damaged in many places by the Turks; that other roads were merely tracks on the side of the hill or up the stony beds of wadis; that throughout the hills the water supply was scanty without development ";* that the transport resources for such operations were limited, as borrowed transport from the 54th Division had now to be returned; that the troops in the vicinity of Junction Station at this time had had severe casualties; and that they were tired after much marching and fighting.

The 52nd Division captured Mansura.

The 22nd Mounted Brigade captured Naane.

The Australian and New Zealand Division gained and held the high ground north of Surafend at Ayun Kara, six miles south of Jaffa, in spite of heavy counter-attacks by the enemy.

November 15*th.*—The Yeomanry Division and part of the Imperial Camel Corps occupied high ground east of Junction Station. The 75th Division attacked from the west a strong position covering the main road from Ramleh to Jerusalem. There the Turkish rearguard had established themselves in the foot-hills at Shusheh, while the 6th Mounted Brigade attacked from the south.

The Turks were driven back to Amwas. The Turks lost 431 killed, 360 prisoners, and 1 gun. At Ludd 300 Turks surrendered to the 1st Australian Light Horse Brigade.

The 54th Division left Gaza in order to join the XXI Corps.

The Turkish detachment from Dhahariye withdrew to Hebron.

November 16*th.*—Jaffa was occupied by the Australian and New Zealand Mounted Division, while the Australian Mounted Division advanced on Latrun.

The orders for the advance on Jerusalem now were for the 75th Division to advance up the Latrun-Jerusalem road, with the 52nd Division on its northern flank marching in tracks three miles north of the main road and the Yeomanry Division covering its left flank. It was hoped that it would be possible to get astride the Jerusalem-Nablus road to isolate Jerusalem and to avoid any fighting there.

November 17*th.*—The yeomanry began to advance towards Bireh through the Ajalon Valley to Berfiliya.

* " Despatches."

The 52nd Division marched to Ludd.

The 54th Division sent one brigade to Yebna.

The 75th Division prepared for their advance on Latrun.

November 18th.—The Australian Mounted Division demonstrated round the south of Latrun. This outflanking movement caused the Turks to evacuate their strong position during the night.

The 8th Mounted Brigade reached Tahta.

The 60th Division concentrated at Gaza.

The 10th and 74th Divisions concentrated north of Belah.

November 19th.—The weather became very wet and cold. There was great difficulty in getting the guns along. There was, however, gradual progress towards the objectives, in spite of enemy opposition in the Saris defile east of Amwas and on our northern flank. The objectives now were (*a*) for the 75th Division with the 5th Australian Mounted Brigade on its southern flank to advance on Bireh via Enab and Biddu; (*b*) for the 52nd Division to support the 75th Division by advancing via Likia and Dukka.

The 60th Division started north from Gaza.

The Australian and New Zealand Division was in touch with the Turks on the River Auja.

November 20th.—Enab, a dominating hill 3,000 feet high, was captured with the bayonet by the 232nd and 233rd Brigades.

The 5th Mounted Brigade advanced in an easterly direction up the Wadi Surar.

The 52nd Division advanced four miles up to Dukka.

The Yeomanry Division reached within half a mile of Beitunie and Ram Allah, but owing to opposition by superior numbers, whose guns outranged the mountain artillery, they had to withdraw to Foka.

November 21st.—The 60th Division arrived at Mejdel.

The 54th Division watched the Ludd-Jaffa line.

Progress towards Jerusalem was slow. The commanding spur four and a half miles north-west of Jerusalem, called Nebi Samwil, was captured by the 75th Division. Kastal Ridge, two miles west of the Shechem-Jerusalem road, was also captured.

November 22nd.—The limit of our advance towards Jerusalem was reached by this day. We had, however, secured the difficult passes through the hills, and, as Lord Allenby wrote : " The troops had achieved valuable results. Had the attempt to force the narrow passes from the plain to the plateau of the Judean range not been made at once, or had it been pressed with less determination, the enemy could have had time to reorganize his defences in the passes lower down, and the conquest of the plateau would have been slow, costly, and precarious. As it was, positions had been won from which the final attack could be prepared and delivered with good prospects of success.''

The Turks' Seventh Army made strong counter-attacks against Nebi Samwil. These we repulsed.

The yeomanry maintained their position at Foka.

The 52nd Division captured Beit Izza, but was unable to take Jib. The enemy, reinforced with fresh and trained assault troops and machine guns, could support their attacks from gun positions on the main road. Our troops could not be adequately supported in this hilly, roadless part of the country.

The 60th Division arrived at Junction Station.

November 23rd.—The 75th Division was unable to capture Jib. The 8th and 22nd Mounted Brigades from Tahta sent back their horses to Ramleh. The 74th Division began to march from Belah to Gaza.

November 24th.—The Australian and New Zealand Division with two companies crossed the River Auja and secured a position north of Hadrah and Muannis.

The 52nd Division again attacked Jib in co-operation with the Yeomanry Division. No progress, however, could be made.

November 25th.—For the next fourteen days there was a pause in our offensive operations. During that period reliefs among the forward troops were carried out. Roads and tracks were improved to carry heavy artillery, reserves of ammunition and supplies were brought up, and the water supply was developed.

The XX Corps was brought into a position from which to co-operate in the final attack.

The Turks, however, constantly attacked throughout this period.

The 3rd Turkish Division regained their positions on the north banks of the River Auja.

Enemy attacks at Nebi Samwil, Beit Izza, near Foka, and Meita were repulsed.

The 74th Division reached Mejdel.

November 26th.—The 60th Division began to relieve the 75th and 52nd Divisions.

November 27th.—Turks advanced in strength towards the River Auja at Mulebbis.

A small detached post of 3 officers and 60 other ranks at Zeitoun was attacked by 600 Turks supported by machine guns and artillery, while other bodies of the enemy tried to work round our flank in this area between the yeomanry and

the 54th Division.   They were unsuccessful in both their
efforts.   The yeomanry post held out till reinforced by the
7th Mounted Brigade from Deiran.

A light A.C. Battery on the Ludd-Sira road stopped the
enemy's advance on Suffa.

November 28th.—The Turks attacked with 3,000 rifles, four
batteries, and mountain guns against the left flank of our
position in the hills between Burj and Foka.   After these
attacks our posts at Foka and Zeitoun were withdrawn.

The 156th Brigade and 4th Australian Light Horse Brigade
reinforced this part of our line.   The enemy's attacks were
then repulsed.   Their attacks against our troops at Suffa
were unsuccessful during the night.   At daylight they suffered
heavily from our artillery fire. and we captured 300 prisoners.

November 29th.—The enemy made a night attack across
the River Auja and gained a footing on the southern bank.
The troops who penetrated our position were outflanked and
captured on the following day.

Attacks against our position at Nebi Samwil were unsuccessful.   At this position the enemy's casualties had been very
heavy.   In four days we took 750 prisoners.

November 30th.—Reliefs were completed.   Our positions
from the Jerusalem-Jaffa road to the sea were held as follows:
60th, with the 10th Australian Light Horse Regiment on their
southern flank, 74th, 52nd, Australian Mounted Division (less
one brigade), 54th Division (less one brigade), Australian and
New Zealand Mounted Division, and 161st Brigade.

The plan for the next offensive movement was for the main
attacks to be from the west and from the south-west of
Jerusalem beginning on December 8th.   There was to be no
fighting close to the city and no injury to any of its buildings.
It was hoped that positions would be gained astride the
Jerusalem-Shechem and the Jerusalem-Jericho roads before the
Turks could escape by either of these main lines of com-
munication.

The 53rd Division was to be employed for the turning movement from the south.

December 1st.—The 53rd Division (less 158th Brigade) began
to march north to Hebron.

Troops of the 19th Turkish Division attacked at Burj
between the 3rd and 4th Australian Light Horse Brigades,
and also at Tahta and at Nebi Samwil.

The enemy lost heavily in all these attacks in casualties
and in prisoners.

December 2nd.—The Yeomanry Division and the 7th
Mounted Brigade arrived at Akir as Corps Reserve.   The 10th
Division relieved the 52nd Division.

December 3rd.—Devon Yeomanry captured Foka. Owing to enemy's counter-attack from ground dominating their flanks the yeomanry were forced to evacuate it.

December 4th.—The 10th Division extended their position east of Dukka.

December 5th.—The 10th Division occupied Suffa and Hellabi. The 74th Division took ground to the south and occupied the Nebi Samwil sector.

December 6th.—The 53rd Division (less 158th Brigade) was at Bilbeh, ten miles north of Hebron. The plan was for these troops to reach a position south of Jerusalem from Bethlehem to Beit Jala on the 7th, and by dawn on the 8th a line three miles south of Jerusalem, when with the 60th, 74th and 10th Divisions and Australian Regiments they would close in on this city from the south, west, and north-west.

The Australian Mounted Division (less one brigade) joined up the line between the 10th Division and the 233rd Brigade of the 75th Division. This line was continued to the sea with positions covering Ludd, Ramleh, and Jaffa by the 54th (less 161st Brigade), 52nd Division, Australian and New Zealand Mounted Division, and 161st Brigade.

December 7th.—The heavy rain and mist made inter-communication and co-operation difficult. The roads were impassable in places for camels and for motor vehicles.

Our attacking divisions were disposed of as follows: 53rd Division, facing north-east, was south-west of Bethlehem, in touch with the 10th Australian Light Horse Regiment connected with the 60th Division south of the Jaffa-Jerusalem road at Ain Karim. The 74th Division, facing east, was north of this road and west of Nebi Samwil. The 10th Division, facing north-east, was between Sira and Beitunia.

During the night the 179th Brigade crossed the Wadi Sudar and occupied high ground south of Ain Karim.

December 8th.—The main assault started at dawn and troops quickly gained their first objective. Owing to the difficulty of obtaining close artillery support in this hilly country, in which it was difficult to move guns forward, the further advance was slow.

By midday, however, the 74th Division had captured Beit Iksa, and the 60th Division had gained a commanding position east of Wadi Sura.

The 53rd Division was delayed by the difficulty of the ground and by the enemy in country favourable for delaying action. Thus the right flank of the 60th Division was exposed when they reached a position about one and a half miles west of Jerusalem, and it was necessary to form a defensive flank.

The 53rd Division later advanced through Bethlehem and Beit Jala.

When the attacks ceased on this day, all the dominating positions west and north-west of the enemy's main line of retreat to the north had been captured.   It had been hoped to gain a position outside this line, but the enemy's opposition had been stubborn and the weather had made progress for men and guns slow in this rocky country.

December 9th.—The Turks made no effort to recover their lost position, and during the night they evacuated Jerusalem. The Mayor of the town brought out a letter of surrender at 0830 hours.

When the advance was resumed, there was no difficulty in driving back the weak Turkish rearguards and in gaining a position four miles north of Jerusalem astride the road to Shechem.

This was accomplished by the 60th Division.   The 53rd Division, advancing north-east, was astride the road to Jericho by 1100 hours, a short time after the bulk of the Turks had passed eastward.   Up to date in these operations 12,000 prisoners and 100 guns had been captured from the Turks.

December 10th.—The XX Corps till December 20th consolidated their position and improved communications.

December 11th.—The Commander-in-Chief made his official entry into Jerusalem.

December 13th.—The 53rd Division advanced north-east of Jerusalem.   The XX Corps now held positions four miles east and north of Jerusalem through Tahta to two miles north-east of Sira.

The Australian Mounted Division (less one brigade) connected up with the 75th Division on the right of the XXI Corps north-east of Ludd.   The line was continued north-west to the sea by the 54th and 52nd Divisions.

The Seventh and Eighth Turkish Armies were now isolated with separate lines of communication twenty miles apart via Nablus and Tul Keram respectively.   The only direct lateral communication between these two lines of supply was the road thirty miles north of Jerusalem joining these two places.

December 14th.—Orders were issued for the advance to an east and west line twelve miles north of Jerusalem, after the necessary improvements had been made in the communications and in the supply services of the XX Corps.

This Corps had six weak Turkish divisions in its front.   The XXI Corps was to put eight miles between the three weak divisions in its front and Jaffa.   This was accomplished by December 22nd by the 52nd Division, after crossing the River Auja.

E

December 15*th*.—The 54th Division occupied Tireh.   The 75th Division occupied Kibbiah and Ibaneth.

December 16*th*.—Preparations were made for the relief of the brigades of the 52nd Division from their positions in the coastal area on the left of our line.   This division was required for the crossing of the River Auja.   Bombardments were carried out nightly, up to and including the night of the crossing, with artillery and machine guns at the same hour and for the same length of time.   This was intended to induce a belief in the minds of the Turks that nothing unusual was happening when we actually crossed.   This ruse was successful.

December 19*th*.—The necessary reliefs and preparations without attracting the enemy's attention had been carried out to enable the 52nd Division to effect a crossing of the River Auja.

The plan for the operations was for the 54th Division to engage the enemy on the right, to advance northward, and to capture Rentia and Fejja; for the 52nd Division, after sending over a covering force to cross with three brigades near the mouth of the river at the Ford, below Muannis and at Stone Bridge, to capture Sheikh Muannis and Khirbet Hadrah.

The Navy was to co-operate by bombarding the Turks' positions at Jelil, Haram, and Arsuf.

December 20*th*.—Owing to complete and careful preparations, precise orders, and adequate arrangements for surprising the enemy, the crossing of the covering party was effected by 2230 hours.   By midnight the 156th Brigade had crossed the river.

December 21*st*.—The assault on the enemy's trenches 1,000 yards north of the Ford began at 0025 hours by the 156th Brigade.

The Turks were completely surprised.   Their main preparations were made for opposing a crossing at Stone Bridge. The 155th Brigade crossed in rafts.   By 0540 hours we had taken Hadrah and Stone Bridge.   By 0200 hours the 157th Brigade had crossed.   Four hours later they had gained all their objectives.

At 0600 hours two batteries crossed the ford.

Eleven officers, 305 prisoners, and 10 machine guns were captured on this day.

December 22*nd*.—The 54th Division carried out their operations as planned for them.   The 52nd Division pushed forward to the high ground at Arsuf, two miles beyond the allotted objective, in order to prevent the enemy from getting observation for their artillery on to Jaffa.   The line reached by this division was from Tel Mukhmar to Arsuf.

The New Zealand Mounted Rifle Brigade crossed the River Auja and captured Tel Nuriyh.

The Royal Navy co-operated in accordance with the plans.

December 23rd.—The 180th Brigade advanced at dawn to capture positions held by the Turks at Adaseh and north of Hannina. They were unable to gain their objectives before the operations were abandoned owing to the weather.

December 26th.—The Turks' XX Corps reinforced from Jericho gained our outpost position on either side of the Jerusalem-Shechem road at 2330 hours in the vicinity of Ras Tawil, east of this road.

The 24th Welch Regiment captured and held a commanding position close to Hill 1910.

December 27th.—The Turks attacked strongly the 53rd and 60th Divisions. The Corps Commander ordered counter-attacks to be delivered by the 74th and 10th Divisions against the right of the enemy's attacking troops at 0630 hours. They made special efforts to gain the Ful Hill from which Jerusalem can be overlooked. On the whole front the enemy's attacks were repulsed, and the effect of our counter-strokes was to draw off the enemy's reserves from the front of the 60th Division, to cause them to lose the initiative and to conform to our offensive operations, and to enable us to advance 4,000 yards on a six-mile front to Sheikh Abdullah, to Jeirut, the Zeitoun Ridge, and Dreihemen.

December 28th.—The 60th Division secured their right flank by the capture of Adaseh. In addition they seized later Jib, Bir Nebala, Er Ram, and Rafat.

The 74th Division captured Hill 250 north-east of Dreihemen and Beitunie.

The 10th Division gained Kefr, Skiyan, Ainein, and Rubin.

The 53rd Division secured Anata and Suffa.

December 29th.—The Australian Mounted Division advanced north-east of Kuddis and Harith. The enemy's opposition was slight. There was a general advance on the front of the XX Corps, whose positions now were from Deir Ibn Obeid to Hizmeh, Jeba, Beitin, Burj, Ras Kerker, to Deir el Kuddis on the left of the 10th Division. On their right was the 74th Division up to the Nablus-Jerusalem road, where they joined with the 60th Division, whose positions were inclusive of Beitin. The 53rd Division on the right carried on the line of positions astride the Jerusalem-Jericho road.

December 30th.—The XX Corps continued to advance and to improve their positions in the vicinity of Beitin, Balua, and Kuddis.

The XX and XXI Corps now satisfactorily covered Jerusalem, Ramleh, and Jaffa, and the road which joined these places.

E 2

In the last three days' fighting the Turks' casualties were over 4,000; 750 prisoners and 24 machine guns had been captured; and an advance of six miles, in the vicinity of Bireh, and three miles from Tahta, had been made; and the enemy had been completely defeated in an attempt to recapture Jerusalem.

It was necessary now to improve roads and to bring the railways nearer to our front. There were, therefore, no further active operations on a large scale until February 19th, 1918. An advance was then made to secure our right flank by driving the enemy beyond the River Jordan as a preliminary to the operations for a further offensive campaign, and to obtain a starting-point for an attack on the Hedjaz line in co-operation with the Arabs based on Aqaba.

This offensive campaign was necessary in order to deal a final blow to Turkey whose limits of endurance were nearly reached, to gain a complete victory after our initial successes, and to compensate for our lack of success in breaking through the German line on the Western Front.

1918.

February 19th.—The advance began to the River Jordan and to the Wadi Auja. The troops employed for this expedition were the 60th, the 53rd, and the Australian and New Zealand Divisions. The natural obstacles were considerable in the seventeen miles of country from Jerusalem to the Jordan. From hills 2,000 feet high the descent to the Jordan Valley, 1,200 feet below sea-level, was very steep in places and broken up by precipitous wadis.

The 53rd and 26th Turkish Divisions in the vicinity of the positions held by our 60th and 53rd Divisions were about 7,000 strong. On the front of these two divisions there was an average advance of three miles.

The 60th Division by 0900 hours captured a dominating height eight miles south-east of Jerusalem. The Australian and New Zealand Division assembled behind this feature with a view to entering the Jordan Valley in the vicinity of Nebi Musa, in order to cut off the enemy's retreat from Jericho.

The 60th Division also took Arak Ibrahim and Ras Tawil. The 53rd Division occupied Ramman.

February 20th.—During the night February 19th/20th, two brigades of the 60th Division moved into the Wadi Sidr, to positions from which they could continue the attack at daylight.

In spite of the enemy's counter-attacks during the night, Talaat Dumm was captured by 0715 hours by the 180th Brigade. The 179th Brigade, supported by artillery, was able to capture Jebel Ekteif, in spite of the difficulties of the country, by midday.

The Australian and New Zealand Division encountered stiff opposition in their advance on Nebi Musa.

The enemy's artillery and machine guns were able to fire on to the most direct approaches that were passable.   The 1st Australian Light Horse Brigade, however, found a way round into the Jordan Plain, and gained a position in the Wadi Jofet Zoben.

February 21st.—The Turks withdrew during the night of February 20th/21st.   The 1st Australian Light Horse Brigade entered Jericho and sent patrols up to Ghoraniyeh Bridge and to the Wadi Auja.

The New Zealand Brigade occupied the Turkish base on the Dead Sea.

The 60th Division advanced to the high ground overlooking Jericho.

The Turks in the Jericho area were now driven to the vicinity of the River Jordan at Ghoraniyeh and the Nahr Auja.

In order to co-operate with Feisal's Arab Army and to interrupt the enemy's communication between Maan and Damascus, it was necessary to extend the area held by our troops.

March 8th.—Our further advance began.   There was little opposition and both corps gained ground.

March 9th.—The 60th Division, after a night march across the Wadi Auja, seized an important position, Abu Tellul, after considerable fighting.   The 53rd Division gained Tel Asur. The 74th Division occupied positions on the ridges overlooking the Wadi Nimrin, and the 10th Division continued the line so as to command the Wadi Jib.   This position they held in spite of repeated counter-attacks.

March 11th.—The 74th Division advanced on to the hills commanding Sinjil.

March 12th.—The 54th and 75th Divisions had heavy fighting before they captured Wadi Ballut and Ras el Ain.   An advance up to seven miles on a twenty-six-mile front had now been carried out by the XX and XXI Corps, who had gained strong positions from which operations could be undertaken east of the River Jordan in conjunction with the Arab army advancing from their base at Aqaba.

The Commander-in-Chief now decided to make a raid on the Hedjaz railway and to destroy the viaduct and tunnel at Amman, thirty miles north-east of Jericho.

March 21st.—The expedition under the G.O.C. 60th Division started off down the Achor Valley to cross the River Jordan at Ghoraniyeh and Hajlah, while feints elsewhere were to be made.   The 180th Brigade was to cross first and was to form bridgeheads at these places.

The troops to take part in this raid were the 60th Division, Australian and New Zealand Division, Camel Brigade, a mountain artillery brigade, a light armoured-car brigade, a heavy battery, and two bridging trains.

March 22nd.—Our attempts to cross at Ghoraniyeh failed, as the current was running at eight miles an hour and the enemy were not surprised.   A battalion of the London Regiment, however, crossed at Hajlah by 0745 hours, and a pontoon bridge was then thrown across the river.   Another bridge was made during the afternoon at this place, and three battalions crossed the river.

March 23rd.—A regiment of the 1st Australian Light Horse Brigade crossed the river and cleared the Turks from Ghoraniyeh.

Three more bridges were built.   This enabled the 60th Division to cross.

March 24th.—The 60th Division occupied Shunet Nimrin (one mile north of Ghoraniyeh) and then marched four miles north-east on Salt.   The mounted troops advanced by tracks south of the Salt-Amman road.

March 25th.—Cavalry reached Naaur and Salt.

March 26th.—Cavalry captured 170 prisoners near Sweileh. A demolition party blew up the railway seven miles south of Amman.

March 27th.—The 181st Brigade (less two battalions) reached a position two miles east of Sweileh.   Demolition parties destroyed the railway at Libban and seven miles north of Amman.

March 28th.—Two battalions of the 181st Brigade, the New Zealand Mounted Brigade, supported by the 9th Mountain Brigade, attacked Amman.   This attack did not progress, owing to the strength of the enemy's position at Amman, and at Hill 3039, owing to the fact that the enemy was not surprised, and owing to the difficulty of artillery support and observation.

March 29th.—Attacks were renewed.   The enemy, 4,000 strong, supported by fifteen guns, were able to hold their positions while their reinforcements tried to envelop the left flank of the 181st Brigade.   They also attacked the 179th Brigade at Salt.

March 30th.—At 0200 hours a night attack was made on Amman.   Part of Hill 3039 and the outskirts of Amman were taken by the New Zealand Brigade, but we were unable to advance farther during the day.   As the Jordan was rising, and as enemy reinforcements were arriving from the north at Amman, and also close to Salt, orders were issued for a withdrawal of the force west of the River Jordan.

March 31*st.*—The withdrawal continued.

April 2*nd.*—The troops were west of the River Jordan, except for those forming a bridgehead at Ghoraniyeh. A total of 953 Turkish prisoners were brought back.

The result of the raid was to draw Turks from the Arab front to defend Amman. The Arabs then cut large sections of the railway line north and south of Maan, and were able to press their attacks on this place.

April 9*th.*—The XXI Corps captured the village of Kefr Ain, Berukin, and Rafat.

April 11*th.*—Enemy counter-attacks were driven off at Berukin and Rafat. The front of twelve miles in this area was advanced to a maximum depth of three miles; and the Wadi Ballut was in our possession. This wadi facilitated inter-communication and water supply.

The Turks unsuccessfully attacked Ghoraniyeh and Mussala-beh. They withdrew to Shunet Nimrin. A raid to Salt in co-operation with the Beni Sakhr Arab tribe began. The force available for this expedition was as follows:

60th Division.
Imperial Service Cavalry and Infantry Brigades.
Desert Mounted Corps (less one division).

April 30*th.*—The 60th Division, with their left flank protected by one brigade, Australian Light Horse, attacked Shunet Nimrin, while the Australian Mounted Divisions captured Salt. The Desert Mounted Corps sent one brigade to a position two miles east of Es Shert.

May 1*st.*—A Turkish cavalry division and part of their 24th Division, having crossed the River Jordan at Damie, attacked the detached mounted brigade, which had to fall back to the Wadi Abyad with the loss of 9 guns.

May 2*nd.*—Two Turkish battalions attacked the cavalry at Salt. The 60th Division was unable to capture Shunet Nimrin. The Arab co-operation did not materialize. The Commander-in-Chief therefore ordered the raiding force to withdraw west of the River Jordan, as the position of the troops at Salt was dangerous.

May 4*th.*—The raiding force was west of the River Jordan. These raids caused the enemy to dissipate their forces, as they were uncertain as to whether our main attacks would in future be on the east or on the west side of the River Jordan.

Therefore they kept their Fourth Army east of the River Jordan; their Seventh and Eighth Armies were west of the river.

April and May.—The Egyptian Expeditionary Force was

organized as an Indian Force. This facilitated supplies, as the Indian Ocean was not liable to attack by German submarines. The 52nd and 74th Divisions and also 24 battalions, 9 yeomanry regiments, and 5½ heavy batteries and 5 machine-gun companies, were replaced by the 3rd and 7th Indian Divisions, 24 Indian battalions, and the 4th and 5th Cavalry Divisions.

June 8th.—Two battalions captured Arsuf.

July 6th.—Maan was invested by the Arab Army, and Medina was isolated by their demolition of the Hedjaz railway north of this place.

July 13th.—A party of guides raided the Turkish trenches in their front and captured 13 prisoners and 1 machine gun.

July 14th.—Three thousand Turks attacked and seized Abu Tellul in the Jordan Valley. The 1st Australian Light Horse Brigade counter-attacked and retook the post with 338 prisoners and 8 machine guns.

Turks also attacked near Hennu Ford from the eastern bank of the Jordan. The Imperial Service Cavalry Brigade charged, dispersed the Turks, and captured 152 prisoners, 2 guns, and 3 machine guns.

July 27th.—A company of the 53rd Sikhs raided the enemy trenches, capturing 33 prisoners and 2 machine guns.

August 8th.—The Arab Army seized Medawera on the Hedjaz railway, sixty-five miles south of Maan. They captured 120 Turks, 2 guns, and 3 machine guns.

August 12th.—Four battalions attacked and surprised the enemy north-west of Sinjil, capturing 239 prisoners and 13 machine guns.

Before offensive operations on a large scale were undertaken, communications were improved and supplies were brought up. Jerusalem was connected to Bireh by railway, and Jaffa was linked up with Ludd and up to a point close to Arsuf. At Jerusalem, Ludd, and Talaat Dumm large reserves of supplies and ordnance material were accumulated.

Water supplies were improved. Roads were widened and metalled. A big bridge suitable for bearing mechanical traffic and heavy guns was built over the Jordan at Ghoraniyeh.

The Turks' three armies now in our front totalled 4,000 sabres, 32,000 rifles, and 400 guns. Our four mounted divisions, seven infantry divisions, and French brigade comprised 12,000 sabres, 57,000 rifles, and 540 guns. The Arabs on our eastern flank were approximately 8,000 strong.

In view of his superiority in numbers and in resources, the Commander-in-Chief decided to take the offensive. The area for these operations was to be west of the River Jordan. The country was more suited for mounted action than in the difficult and hilly country east of the River Jordan. The important positions on the Turks' lines of communication, which in our possession would block the enemy's line, were Afule, Beisan, and Deraa. These could be more quickly reached by mounted troops acting in the coastal area than in the hilly country. Therefore, the Commander-in-Chief decided to break through the enemy's positions on their right flank, so that the cavalry could arrive at Afule and Beisan more quickly than their Seventh and Eighth Armies could. Later their Fourth Army would be similarly cut off by our occupation of Deraa.

In order to carry out these plans it was necessary to concentrate in the coastal area five divisions (35,000 rifles), 383 guns, and three mounted divisions, and at the same time to keep the enemy apprehensive of an attack on his left flank.

Our air supremacy was the main factor in making possible the secret concentration on our left flank, involving as it did a withdrawal of troops from the remainder of the front.

September 9th.—Force Order No. 68 to the Desert Mounted Corps was issued. The gist of this order was as follows:

The XXI Corps was to make a gap in the enemy's positions in the coastal area. The Desert Mounted Corps was to pass through it and across the Nahr Falik and the Khez Zerkiyeh, then it was to march as early as possible to reach Afule and Beisan before the enemy could assemble troops or material there.

Roads and railways through Jenin were to be blocked, and the bridge over the Jordan at Mejamie was to be seized.

September 16th.—The Arab Army destroyed a bridge and part of the railway line fifteen miles south of Deraa. The Royal Air Force damaged the station and line at this place.

September 17th.—The Arab Army destroyed the railway line north and west of Deraa.

September 18th.—Concentration of infantry and guns required for offensive operations in the coastal area was complete. The 4th and 5th Cavalry Divisions were in the vicinity of Sarona. The Australian Mounted Division was at Ludd.

September 19th.—The main attack, starting at 0430 hours, was preceded by a fifteen-minute bombardment against the front held by the three Turkish divisions, 8,000 strong, with 1,200 sabres and 130 guns from Rafat across the Plain of Sharon to the sea. Pivoting on the French contingent, which advanced north of the Wadi Ballut, by 0800 hours our divisions penetrated the enemy defences to a depth of five miles in a north-easterly direction, and then advanced east through Hable Kalkilieh, Tireh, to Tayibieh and Tulkeram in the following order from the south: 54th, 3rd, 7th, 60th Divisions.

The 75th Division after capturing Tireh remained there in Corps Reserve.

By 1100 hours this wheel to the east had driven back the bulk of the Eighth Turkish Army, and enabled the main body of the cavalry to advance up the Plain of Esdraelon; one Australian regiment reached a position during the evening four miles east of the 60th Division at Tulkeram.

The 5th Cavalry Division covered twenty miles up the coast to the vicinity of Hudeira, helped by the guns of the Royal Navy.

The Messudie defile and the Bireh-Nablus road became blocked with troops and transport, and were heavily bombed by the Royal Air Force.

The XX Corps made a night attack. The 53rd Division attacked east of the Nablus-Bireh road, and the 10th Division advanced west of this road in a north-easterly direction.

Progress was made by both divisions, in spite of the difficulty of the country and the strength of the enemy's organized positions.

September 20th.—The 10th Division advanced seven miles from its original line. The 53rd Division captured Kefr Malik and maintained their positions in this area, in spite of heavy counter-attacks.

The XXI Corps continued to advance in an easterly direction and reached the line through Azzon, Kefa-Lukir, Beit Lid, Messudie Station. The R.A.F. bombed the Messudie-Jenin road, as it was crowded with troops and transport.

The 4th Cavalry Division reached Beisan by 1630 hours, and the 19th Lancers secured Mejamie Bridge. The 5th Cavalry Division captured Afule and Nazareth.

The Australian Mounted Division occupied Samaria and Jenin.

" Thus within thirty-six hours of the commencement of the battle all the main outlets of escape remaining to the Turkish Seventh and Eighth Armies had been closed." (Despatches.)

September 21st.—The XX Corps captured Nablus with the co-operation of the 5th Australian Light Horse Brigade and the French cavalry.

The 60th, 7th and 3rd Divisions reached positions covering the road and railway at Samaria, Messudie, and Tulkeram.

Cavalry collected a large number of prisoners during the day.

Chaytor's Force on our right flank in the Jordan Valley secured the flank of the army and kept in touch with the enemy in this area.

The R.A.F. bombed the enemy's columns and transport along Wadi Farah to Damie.

September 22nd.—The bridge at Jisr Damie was seized by

the New Zealand Mounted Brigade and the 1st West Indies Regiment of Chaytor's Force. The west bank of the River Jordan was now cleared of the enemy.

The 11th and 12th Armoured Car Batteries reconnoitred towards Haifa. The 5th Cavalry Division concentrated towards Nazareth.

The R.A.F. bombed enemy columns on the Ferweh-Beisan road.

September 23rd.—The 13th Cavalry Brigade occupied Acre. The New Zealand Mounted Brigade occupied Salt.

The 15th Imperial Service Brigade captured Haifa.

The 11th Cavalry Brigade seized a ford five miles south of Beisan and prevented a large body of Turks from crossing. These were captured; those who had crossed were dealt with by the 29th Lancers and Middlesex Yeomanry. Maan was evacuated by the Turks and was entered by the Arabs.

September 24th.—The New Zealand Mounted Brigade passed through Suweileh. The 4th Australian Light Horse Brigade captured Semakh.

The Arab Army was fourteen miles south-east of Deraa, which was bombed by the R.A.F. The Fourth Turkish Army was now retreating and was pursued by the R.A.F. with low-flying action.

September 25th.—Tiberias was occupied.

Chaytor's Force captured Amman, with 5,000 prisoners. The Turks who escaped in a northerly direction were harassed by the Arabs. Orders were issued for the Desert Mounted Corps advancing in two columns, one on either side of the Sea of Galilee, to occupy Damascus.

September 26th.—Chaytor's Force remained at Amman. The Arab Army moved to Sheikh Saad, north-west of Deraa. The 10th Cavalry Brigade occupied Irbid after much fighting.

September 27th.—The Australian Mounted Division, marching north of the Lake of Tiberias, met with opposition at the Benat Yakub Bridge, later at Kuneitra and Sasa, twenty miles from Kuneitra.

The enemy's position at the bridge was turned by fording the river south of the bridge.

Haifa was opened as a port for supplies.

The 10th Cavalry Brigade, marching south of the Lake of Tiberias, occupied Remte.

The Arab Army captured Deraa.

September 28th.—The 4th Cavalry Division and the Arab Army met at Remte.

September 29th.—The Maan garrison with 12 guns and 35 machine guns surrendered to the 2nd Australian Light Horse Brigade near Kastal.

The 3rd Australian Brigade attacked Sasa.

September 30th.—Sasa was captured by the 12th and 4th Australian Light Horse Regiments.   The 13th Brigade captured Kiswe.   The 5th Cavalry Division captured Kaukab.

The 3rd, 4th and 5th Australian Light Horse Brigades were astride the exits from Damascus on the north-west and north-east.

The 14th Cavalry Brigade reached Kadem Station.

October 1st.—The 14th Cavalry Brigade and the Arab Army entered Damascus.

The 3rd Australian Light Horse Brigade blocked the northern exits of the town and attacked the enemy at Duma.

The 10th Australian Light Horse Regiment continued the pursuit of the Turks up to Khan Kusseir.

The 7th Indian Division from Haifa left for their march to Beirut.

The 5th Division concentrated at Damascus.

October 2nd.—The 3rd Australian Light Horse Brigade pursued a column escaping north-east from Damascus for six miles, then charged the Turks and captured 1,500 prisoners, 3 guns, and 26 machine guns.

October 6th.—Rayak Junction and Zahle were occupied by the 4th and 5th Cavalry Divisions.   A total of 177 prisoners and 5 guns were captured.

October 8th.—Ships of the French Navy entered Beirut harbour.   The 7th Division reached Beirut.

October 11th.—The 13th Brigade occupied Baalbek.

October 13th.—The 5th Cavalry Division, Column " A," consisting of Divisional Headquarters, two regiments of the 15th Brigade, armoured-car batteries and 3 light-car patrols, reached Lebwe.

The XXI Corps cavalry and armoured cars occupied Tripoli.

October 14th.—Column " A " reached Kaa.   The 19th Infantry Brigade left Beirut for Tripoli, which now became a base for supplies.

October 15th.—Column " A " reached Khan Kusseir.

October 16th.—Column " A " reached Homs.   Reports were received that Aleppo, 120 miles farther north, was occupied by 20,000 Turks.

October 17th.—Column " B," consisting of the remainder of the 5th Cavalry Division, joined Column " A."

October 18th.—The 7th Division reached Tripoli.

October 21st.—Column " A " reached Zor Defai.

October 22nd.—Enemy rearguards were driven from their positions near Khan Sabil.

Column " A " reached Shaikhun.

October 23rd.—Armoured cars engaged enemy cavalry near Maarit Nabnam, which was reached later by Column " A."

October 24th.—Armoured cars engaged the enemy at Khan Tuman. Column "A" reached Seraikin.

October 25th.—Column "A" reached Turmanin. Armoured cars engaged the enemy five miles south of Aleppo.

October 26th.—The Arab Army entered Aleppo.   The 15th Cavalry Brigade and armoured cars reached a position west of Aleppo astride the Alexandretta-Katma road.

Turkish rearguards withdrew to a position twenty miles north-west of Aleppo.

October 28th.—The 14th Brigade occupied Muslimie Junction.

October 31st.—The Armistice was completed with Turkey.

" Between 19th September and 26th October, 75,000 prisoners were captured.   In addition 360 guns fell into our hands, and the transport and equipment of 3 Turkish armies. The 5th Cavalry Division covered 500 miles and captured over 11,000 prisoners and 52 guns."   (Despatches.)

CHAPTER II.

ILLUSTRATIONS OF THE PRINCIPLES OF WAR.

I. SURPRISE.

    1. The plan for and operations of September 19th, 1918.
    2. Raid on August 12th, 1918, to the Burj-Gurabeh Ridge.
    3. Raids on July 13th and 27th, 1918.
    4. Raid on July 6th, 1916, at Aqaba.
    5. Unsuccessful raid on Amman, March 30th, 1918.
    6. Unsuccessful Turkish attack on our bridgehead at Ghoraniyeh, April
        11th, 1918.
    7. Raid on Salt, April 29th, 1918.
    8. Crossing of the River Auja, night December 20th/21st, 1917.
    9. Attack on Beersheba, October 31st, and against the Hureira posi-
        tion on November 6th, 1917.
   10. Charge at Huj on November 8th, 1917.
   11. Actions at Qatiya, April 23rd, 1916.

II. MAINTENANCE OF THE AIM IN WAR.

    1. Force Order No. 68, dated 9/9/18.
    2. Operations after the capture of Beersheba, November 1st to 6th,
        1917.
    3. Operations after the capture of Junction Station, November 14th,
        1917.
    4. Turkish forces in the Atawineh Section, November 7th, 1917.
    5. First attack by the Turks on the Suez Canal, February 1st to 10th,
        1915.
    6. Battle of Romani, August 3rd, 1916.
    7. First Battle of Gaza, March 26th, 1917.
    8. Second Battle of Gaza, April 19th, 1917.

III. ECONOMY OF FORCE.

    1. Original Turkish dispositions.
    2. Turkish detachments at Amman, Salt, and Shunet Nimrin caused
        a dissipation of their forces.
    3. Separation of the Seventh and Eighth Turkish Armies after Novem-
        ber 14th, 1917.
    4. Turks dissipated their reserves in the initial operations up to
        November 6th, 1917.

IV. CO-OPERATION.

    1. The original plan for our operations beginning on September 19th,
        1918.
    2. Operations in the Jordan Valley, February 19th to 21st, 1918.
    3. Lack of co-operation between the Seventh and Eighth Turkish
        Armies after November 14th, 1917.
    4. Co-operation of all arms for the attack on the Turks' main position
        covering Sheria and Hureira, November 6th, 1917.
    5. The capture of Beersheba.

V. Concentration.

    1. Concentration of superior numbers in the coastal area by September 18th, 1918.
    2. Final attack on Jerusalem.
    3. Para. 5, Force Order No. 54.

VI. Security.

    1. Arrangements for September 19th to 21st, 1918.
    2. Failure at Amman, March 28th to 31st, 1918.
    3. Capture of Jericho and positions in the lower Jordan Valley.
    4. Turks' attempts to recapture Jerusalem, December 26th to 30th, 1917.
    5. Reports by the R.A.F.
    6. Turkish positions from Gaza to Beersheba.

VII. Offensive Action.

    1. Turks adopted a defensive policy.
    2. Our operations, September 19th to October 31st, 1918.
    3. Necessity for cessation of offensive operations after the capture of Jerusalem on December 9th, 1917.
    4. Orders by G.H.Q. for November 9th, 1917.
    5. Capture of Beersheba, October 31st, 1917.
    6. Action at Magdhaba, December 23rd, 1916.
    7. Action at Rafa, January 9th, 1917.

VIII. Mobility.

    1. Pursuit of the Turks by the D.M.C. after September 19th, 1918.
    2. Administrative arrangements made to add to the mobility of the E.E.F.
    3. Cavalry operations in the Plain of Sharon, September 19th to 22nd, 1918.
    4. Raid on Salt, April 30th, 1918.
    5. Capture of Katrah and Mughar on November 13th, 1917.
    6. The enveloping movement for the capture of the Gaza-Beersheba line, October 31st to November 7th, 1917.

I. Surprise is the most effective and powerful weapon in war.

(1) The basis of the Commander-in-Chief's final plan for the defeat of the Turks, starting at dawn on September 19th, 1918, was to mystify and mislead them as to the weight and direction of the attack.

In the coastal area, by the concentration of three mounted divisions, five infantry divisions, and 383 guns, unknown to the enemy, a strategical surprise was obtained.

Two raids east of the Jordan had given the enemy the impression that our main attack would be in this area. This impression was confirmed by the activities and demonstrations of General Chaytor's Force in the Jordan Valley. The mounted troops marched from the Jordan by night. Their camps were left standing. On arrival at the coastal area they were concealed in the olive woods and orange groves. The Australian Mounted Division was bivouacked round Ramleh, the 4th Cavalry Division was near Selmeh, and the 5th Division was north-west of Sarona.

The R.A.F. enabled General Allenby to conceal the concentration of his cavalry. Elaborate arrangements were made to mislead the enemy as to the actual point of our main attack. The H.Q. Camp was left standing at Talaat Dumm after the Commander and Staff were established in the coastal area. Dummy camps and horse lines were erected and small parties simulated the activities of grazing guards.

British West Indian battalions, observed by the Turks from Shunet Nimrin, maintained normal camp activities. Battalions marched to these H.Q. down the Jerusalem-Jericho road by day. New bridges were thrown over the Jordan.

Preparations were made for an advanced report centre for G.H.Q. at Jerusalem as if to be in touch with operations on the eastern flank.

The only written orders, issued as late as possible, were by G.H.Q. and the D.M.C. and the XX and XXI Corps.

Nablus telegraph office was heavily bombed on September 18th, so that on the following day Turkish H.Q. were in ignorance as to the extent of our advance. In this way the Commander-in-Chief obtained a complete strategical surprise.

The Turks expected our main attack to be in conjunction with the Arab forces who were in the vicinity of Kasr el Azrak, towards Damascus via Deraa.

On September 19th a complete tactical surprise was obtained when our XXI Corps, 5th Australian Light Horse Brigade, and 60th Division, and a French detachment of infantry and artillery, supported by heavy guns and two brigades of pack artillery, not only broke through the Turks' defences on a front of 14,000 yards to a depth of 3,000 yards, but were able to continue to advance through their second line of trenches 3,000 yards in rear, and then, pivoting on the French troops at Jiljulie, to swing to the east through Hable up to Tulkeram, while the 5th Australian Light Horse Brigade covered the left flank.

The three divisions of the D.M.C. were thus able to advance rapidly up the plain via Mughair and Mukhalid, and by the evening of September 19th had broken through the enemy's defences north and south of the Mefjir River, and had continued the infantry and 5th Australian Light Horse Brigade positions in a north-westerly direction to the coast south of Cæsarea.

(2) On August 12th, 1918, three battalions raided the Turks holding the Burj-Gurabeh Ridge. They had to cover a

distance of 2,000 yards and then climb 900 feet. They were able to envelop the enemy's flanks on a ridge 5,000 yards long held by 800 rifles and thirty-six machine guns. They completely surprised the Turks, and captured 239 prisoners and thirteen machine guns.

(3) On July 13th and 27th, 1918, similar daylight raids were carried out by the Indian troops. In each case prisoners and machine guns were captured.

(4) Aqaba was captured by the Arabs by means of a surprise attack on July 6th. The Turks holding this town expected an attack from the Gulf of Aqaba; instead the Arabs rushed the posts north of Aqaba from the vicinity of Maan, and finally surprised and captured this town from the same direction.

(5) The trans-Jordan raids misled the Turks as to the direction of our main offensive, which started on September 19th, 1918. The first raid on Amman, however, was not a tactical surprise at this place, owing to the heavy and continuous rains adding so much to the difficulties of the country that progress was slow.

The crossing of the River Jordan by the 180th Brigade, supported by the divisional artillery, surprised the enemy in the initial stage of this operation. Feints were made along the river and parties were landed at the same time in motor boats on the north of the Dead Sea to mislead the enemy as to the point where the crossing was to take place, while the 180th Brigade, in spite of heavy rains and a swift current, was able to cross the Jordan at Hajlah unknown to the Turks. A bridge here was constructed early on March 22nd.

The raid operations started on March 21st, but it was not until the early morning of the 30th that the actual attack against the Turkish positions at Amman started.

By this time enemy reinforcements had arrived, and the positions round the town and on Hill 3039 south of it had been strengthened. The Turks also, on the 29th, with two infantry brigades and a cavalry division, attacked the cavalry brigade and infantry brigade at Salt until the night of the 31st, when they withdrew. The result was that there were no reinforcements available for the attacks on Amman, which failed in the first instance owing to the fact that the enemy were not surprised.

We had had 1,600 casualties, the River Jordan had risen 9 feet during the twelve days' rain, the reserves were used up, and the troops were exhausted.

It was, therefore, decided by General Shea to withdraw from Amman on the night of April 30th/31st, although we had not gained all our objectives.

F

We had, however, drawn pressure off the Arab Army in the vicinity of Maan, as the Turks doubled their garrisons at Amman. We inflicted heavy casualties on the Turks and captured 1,000 prisoners and much ammunition at Salt. The belief of our intentions to make our first main attack via Deraa on Damascus was confirmed.

All troops, except those holding a bridgehead west of the Jordan at Ghoraniyeh, were back by April 2nd west of the River Jordan.

(6) On April 11th the Turks made an attack on our positions at Ghoraniyeh bridgehead held by the 1st Australian Light Horse Brigade, and also at the same time against our positions held by the 2nd Australian Light Horse Brigade and Camel Corps west of the River Jordan and north of the Wadi Auja.

Our troops were not surprised. The enemy were driven off at all points, with a loss of 500 killed and 100 prisoners.

(7) In the second raid, carried out east of the River Jordan by two brigades of the 60th Division and the Anzac and Australian Mounted Divisions and the Imperial Service Brigades, the Turks were surprised at Salt, which was captured by the 3rd Australian Light Horse Brigade.

The mounted charge made by the 8th Regiment, after the 9th and 10th Regiments had captured the works covering the town on the north-west, demoralized the Turks, and by 1830 hours on April 29th the town was in our possession.

The Headquarters of the Fourth Army only just escaped. Though Shunet Nimrin was not captured, owing to the difficulties of the ground and the strength of the opposition, and as the Arab co-operation did not materialize, yet the strategical surprise obtained on September 19th was helped by this operation, as the Turks continued to disperse their forces, leaving the Fourth Army east of the River Jordan in anticipation of an attack in force in this area.

(8) The 52nd Division completely surprised the enemy in their crossing of the River Auja on the night of December 20th/21st, 1917. Every evening our artillery shelled the Turkish position on the north bank, so that on the evening of December 20th, after the usual bombardment, the Turks expected no further operations. The result was that our covering parties of the 7th Highland Light Infantry of the 157th Brigade crossed unperceived by the enemy at 2000 hours, 700 yards from the sea.

The 52nd Division then crossed in three brigade columns. The 156th Brigade crossed south of Muannis, the 157th forded the river south of Tel Rekkit, and the 155th Brigade crossed between Jerisheh and Stone Bridge after demonstrating 300 yards east of this bridge.

When the assault started at 0025 hours, all objectives were secured from Hadrah to Tel Rekkit before dawn on December 21st.

(9) The Turks were completely surprised by our attack on Beersheba. It was necessary to gain this position by a surprise before the enemy could reinforce. It was necessary to have it in our possession, so that we could gain ground for manœuvre for our main attack, and also to obtain the water in Beersheba.

This surprise was gained by preparations designed to mislead the enemy.

By the active trench warfare in front of Gaza the Turks believed that our main attack would be against this position. Four days before the attack on Beersheba there was a continuous day and night bombardment on Gaza.

All movements of the XX Corps and D.M.C. to areas of concentration east of Shellal took place at night. Vacated camps were left standing, and the usual routine was continued in them by small parties left behind. Misleading wireless messages were sent to the Turkish stations.   .

Finally, when two divisions of the D.M.C., after night marches from Khalasa and Asluj of twenty-five and thirty miles respectively, reached their allotted positions east of Beersheba by 0800 hours on the morning of October 31st, the Turks were completely surprised.

Most of their troops in the observation trenches were facing south-west. The Turks had then no time to reinforce their detached post at Beersheba or to demolish the water-plant and the seventeen wells.

The Turks were still kept in uncertainty as to where our main blow was to be. The bombardment in conjunction with the Navy was continued at Gaza; and while the XX Corps was preparing for the main attack against the Sheria-Hureira defences, the XXI Corps on November 1st captured the Umbrella Hill-Shaikh Hasan line.

This confirmed the Turks in their belief that the main objective for our attack was to be Gaza. They did not reinforce the Kauwukah defences. They were surprised at dawn on November 6th, when the 10th, 60th, and 74th Divisions successfully attacked their Hureira positions.

(10) During the afternoon of November 8th the 60th Division supporting the Australian Mounted Division was checked in its advance on Huj by a Turkish rearguard of two battalions supported by artillery and machine guns. The troops of the Warwick and Worcester Yeomanry, who were in the

vicinity, made excellent use of the ground to gain a position unseen by the Turks about half a mile from them.

Here they rallied, formed into three lines, and charged. The Turks were completely surprised, and their position was captured. Our losses were 75 out of the 170 who took part in the charge.

(11) The value of surprise was illustrated in the actions on April 23rd, 1916, when the Turks, estimated at 3,650 men, six guns, four machine guns, and six companies of cavalry, made a sudden raid in force upon Qatiya and upon our post at Oghratina. The Turks captured both these places, and also three and a half squadrons of yeomanry and details were lost to the E.E.F.

The Turks, after a night march, reached the vicinity of Oghratina about 0400 hours on April 23rd.

The first intimation of the proximity of the enemy was the sound of pumps being used at wells 500 yards from the posts. At 0575 hours our post, in which were two squadrons (less one troop) of Worcestershire Yeomanry and an R.E. detachment, was attacked on three sides by Turks who had reached a position not more than fifty yards from our forward defences. The morning fog had helped the Turks, as it had prevented adequate morning reconnaissance. The Yeomanry were surprised and overwhelmed after a gallant resistance of three hours, by which time they had suffered 146 casualties.

The Turks then advanced to Qatiya against our isolated detachment, consisting of a squadron of Gloucestershire Yeomanry, a machine-gun detachment, and dismounted details. They attacked at 0800 hours. Later they were joined by a squadron from Hamisah. It was not until 1500 hours that they were forced to surrender. The Turks were helped in these surprise attacks by the fact that our advanced posts were seven and a half miles apart, that the headquarters of the 5th Mounted Brigade were at Romani, and that a proportion of these headquarters was raiding the enemy's empty camps at Mageibra. The Turks in their attacks at Dueidar and Hill 70 were unsuccessful because there was no surprise. At Dueidar there were 156 rifles. At Hill 70 there was one battalion. The Turks started their attacks at Dueidar at 0500 hours, and they were met with such accurate fire that they withdrew after suffering about twenty casualties. They finally retired at 1300 hours, leaving seventy dead and twenty-five prisoners in our hands. Their pursuit and subsequent rout was carried out by the 5th Australian Light Horse Brigade and by the Royal Flying Corps.

II. MAINTENANCE of the aim in war will be the destruction of the enemy's main field forces.

(1) The instructions issued by the Commander-in-Chief to the G.O.C., D.M.C., with reference to Force Order No. 68, dated 9/9/'18, exemplify this principle.

They show that the Commander-in-Chief wished every available mounted man to be placed where he could best carry out the plan of defeating decisively the Turkish Seventh and Eighth Armies. On no account were the D.M.C. to be diverted from carrying out this main objective.

The instructions were as follows:—

"The G.O.C. XXI Corps has been entrusted with the task of breaking down the enemy's resistance in the coastal plain and opening the way for your Corps to cross the Nahr Falik and the Khez Zerkiyeh Marsh. You must on no account allow your troops to be drawn into the infantry fight south of the Nahr Falik, nor, after the passage of this stream, to be diverted from your objective by the presence of hostile troops in the Tulkeram-Kulunsaweh-Tireh area, which will be dealt with by the XXI Corps. You will be responsible for arranging that the line Khez Zerkiyeh-Mouth of the Nahr Falik is crossed at the earliest possible moment. Should any portion of the enemy's force retreat in the direction of Haifa, you will detach only sufficient troops to keep touch with it and protect your line of communications, as it is vital that as large a proportion of your force as possible should be available to carry out the rôle assigned to you, which is to place your troops about Afule and Beisan, where the enemy's railway communications can be cut at their most vital point, and whence you will be in a position to strike his columns if they endeavour to escape in a northerly or north-easterly direction."

The instructions to the XXI Corps (3rd, 7th, 54th, 60th, and 75th Divisions with a French contingent, 5th Australian Light Horse Brigade, two brigades mountain artillery, and 18 heavy batteries) were to enable the cavalry to carry out their instructions as early as possible by breaking through the enemy's defences between Jiljulie and the sea. Then to pivot on this place with the French troops and to advance in an easterly direction on to the line Hable-Kalkilieh-Tulkeram.

The D.M.C., when the XXI Corps was north of the Nahr Falik, was to seize Afule and Beisan to cut off the Turks' retreat at these places, sending only detachments to Nazareth and Jisr Mejamie.

The XX Corps was to contain the Turks on his front, and to block the exits to the lower Jordan Valley.

The objective was maintained until the Seventh, Eighth, and Fourth Turkish Armies had ceased to exist, and a cavalry division was at Aleppo astride the Turkish communications between Constantinople and Baghdad.

(2) On November 2nd, after the capture of Beersheba, the Anzac Division occupied a line to protect this town from the north from Nettar and Hawa to Likia. The Camel Corps Brigade connected up the line with the positions held by the 53rd Division at Toweil Abu Jerwal. On November 3rd there was a general advance. The 53rd Division attacked the Turks at Khuweilfeh and the Camel Corps and Cavalry advanced on Dhahariye. The enemy now made determined attacks during the next three days and nights. They brought in their principal reserves in an endeavour to induce the Commander-in-Chief to alter his plans and to weaken his striking force by bringing reserves to meet these attacks.

The Commander-in-Chief, however, maintained his objective, relying on the troops north of Beersheba to deal with the enemy while he made no essential modification in his plan. Had he conformed to the enemy's operation, his striking force might have been too weak for the attack on the Turks' main position, and it would have led to fighting in hilly country which would have been slow and costly.

The result of the enemy's attacks was that they had heavy casualties, they prolonged their line considerably to the east, and used up reserves. The Commander-in-Chief, on the other hand, had his striking force available for the attack at the decisive time and place at dawn on November 6th. By 1430 hours on this day the whole of the Turks' Kauwukah and Rushdi systems were in our possession.

(3) After the capture of Junction Station on the morning of November 14th, the Commander-in-Chief had a difficult problem to solve. Two Turkish armies were now separated. The cavalry in the vicinity had been continuously engaged since October 31st. In nine days the 52nd Division had marched sixty-nine miles, and the 75th Division had marched forty miles.

Our objective was Jerusalem, where the bulk of the Turkish field army was.

But to reach this place it would be necessary to cross very difficult country in which there was only one road, which was blocked and flanked by strong positions. There was a scanty water supply, and little was known about the country north and south of the main road.

Supply and transport difficulties would be increased if an advance was continued in this area. The Commander-in-Chief, however, determined to continue his advance towards

Jerusalem with the Yeomanry Division and the 52nd and 75th Divisions. By November 24th these troops had gained important positions overlooking Jerusalem and the main Jerusalem-Nablus Road. From these positions success was assured when reinforcements from Hebron and Gaza arrived to take part in the final attack which led to the fall of Jerusalem, and defeat of the Seventh Turkish Army by December 9th.

(4) The Commander of the Turkish forces in the Atawineh sector of their defences maintained his objective by holding his position there until the night of November 7th/8th. By the evening of November 7th the bulk of their Eighth Army was north of Wadi Hesi. Our 52nd Division (less one brigade) and Imperial Service Brigade were north-west of Gaza and south of this wadi. The 60th Division and Australian and New Zealand Divisions were north-east of Atawineh.

The Turkish 26th and 54th Divisions south-east and north-west of Atawineh were never actually attacked. They, therefore, remained in their positions and shelled the right flank of the 54th Division in Gaza, and delayed the advance of the transport being sent to the XXI Corps for the pursuit up the coast. This corps was consequently somewhat delayed.

These troops at and near Atawineh were able to rejoin the Eighth Army during the night of November 7th/8th, before the D.M.C. had joined up with the Imperial Service Cavalry Brigade advancing on the east of the 52nd Division.

(5) In the attack by the Turks on the Suez Canal between February 1st and 10th, 1915, this principle was not carried out. Our initial distribution gave the enemy every opportunity of maintaining the offensive at the point where they would have the best prospects of success. The Turkish Commander, Djemal Pasha, decided to make his main attack with six battalions and heavy artillery against the central sector of our defences between Deversoir and Ferdan. Having made this plan, Djemal Pasha might have gained contact with the troops defending the canal, and he might then have pressed his attack vigorously with his available reserve in the centre of our position. It was here that the sweet water canal was nearest to the Suez Canal, and along the sweet water canal was the best route to Cairo. The capture of Cairo would have enabled Djemal Pasha to carry out his mission of securing Egypt, and of gaining the adherance of the Mohammedan section of the population to the Turkish cause.

He could not with inferior numbers hope to succeed if he split up his force into three parts with no inter-communication between them. He could not hope to have superior numbers at any one point of attack as we had 70,000 men in Egypt at the time, and his striking force was not more than 15,000 strong. He could not expect to surprise our forces as he had to cross the Sinai Peninsula, at least 120 miles wide, before he could get in touch with our troops on the Canal. Our mounted patrols had been in touch with his patrols near Dueidar on January 22nd.

Four days later his troops were located at Mabeuik, Moiya Harab, and Mukhsheib, and near Qantara his troops had failed in their attack to surprise our outposts.

Similarly, his attacks on our posts near and at Kubri, at 0300 hours on January 27th, had been unsuccessful. The situation, however, on February 2nd was that 5,000 Turkish infantry and guns were east of Serapeum, a similar force was outside Ferry Post, and 2,000 Turkish infantry and guns were near Qantara. Thus the bulk of the Turkish troops was split up on a front of 30 miles with no lateral road between these forces. It was probably necessary for the maintenance of his army to march by different routes, but it was unsound policy to split them up and attack at points where he must be inferior in numbers against an enemy strongly entrenched with superior covering fire and prepared for this attack for at least three weeks. Finally, when his attacks were repulsed, Djemal Pasha retired without using his reserve. There had been little vigour in the execution of the attack. On the night February 2nd/3rd, the Turks made weak attempts against our defences at Qantara, Kubri, and Ferdan, and attempted to cross the Canal between Serapeum and Tussum. Of their three pontoons that did cross the Canal, all the occupants were either killed or captured. During daylight, on the morning of February 3rd, they tried again to cross at Serapeum, but they were unsuccessful. Owing to the fire from the guns of H.M.S. *Clio* silencing their artillery at Ferdan, they did not press their attacks in this quarter.

By 1400 hours on February 3rd their artillery ceased fire, and soon the whole force began to retreat by the way they had come. Though their casualties were only 2,000, their morale and prestige suffered, and they completely failed to carry out their objective. Certainly, Djemal Pasha had a difficult task to perform. He would be attacking a force that was more than four times as strong as his own, and though our Canal defence force was divided up into three sectors along the western bank of the Canal, except for three half-battalion posts at Tussum, Serapeum, and Deversoir, and small posts at Ferdan and Kubri, yet we had means of reinforcement with a lateral railway from our Canal reserve at Moascar, and for our further reinforcement from Cairo by the railway from this place to Ismailia. Djemal Pasha's only chance lay in a determined maintenance of his objective by a vigorous attack with all available strength at the point where he hoped to make a crossing.

The plan which he carried out for his raid on the Canal might have led to a complete disaster if our available reserves had been ready to counter-attack and to pursue.

(6) A violation of this principle of maintenance of the objective must be noted again in Kress von Kressenstein's raid on August 3rd/4th, 1916, against our troops covering railhead at Romani, between Hod el Enna and Mahamdiyeh. The Commander of the Turkish troops had forces, approximately 18,000 strong, including 15,000 rifles, 30 guns, and 38 machine guns. Our position was held by the 52nd Division and one brigade of the 53rd Division. Their plan was to attack our position, held by three brigades, from the east towards Romani, and at the same time to endeavour to make an outflanking movement on the south towards Mount Royston against our right flank, where our mounted troops held Wellington Ridge parallel to the railway running between Romani and Pelusium. The difficulty for von Kressenstein was that we had available approximately 30,000 men, namely, two divisions, six mounted brigades, and nine batteries. He had not the superior numbers required for a successful envelopment against troops holding entrenched positions. His plan could only have succeeded if he had surprised us and had tenaciously followed up his initial success. No surprise was possible, as our mounted troops had located their positions on August 2nd, 1916, east of Katib Gannet Hill and Etmaler. They did, however, gain an initial success.

Soon after midnight on August 3rd/4th they advanced, supported by artillery, from Abu Hamra. At first their troops making the frontal attack gained ground. On our right flank the mounted troops were at first pressed back to and North of Mount Royston. By the early afternoon of August 4th the Turks had reached a point one and a half miles from the railway, and along the southern slopes of Wellington Ridge from the north of Mount Royston. However, their frontal and flank attacks were not synchronized, and were not in touch with each other. Nor did the Commander use his reserves to press forward where he had been successful. When our counter-attack from the direction of Pelusium to Mount Royal was successful, the Turks on their left flank withdrew in disorder. Only the difficulties of water supply, intense heat, and the exhaustion to the troops caused by marching in soft sand saved the Turks from complete disaster. The Turks however, lost 50 per cent. of their forces and undertook no further offensive operations.

(7) The difficulty of maintaining the objective was shown at the two battles of Gaza in 1917.

At the first battle of Gaza, March 26th, 1917, we had the advantage of having superior numbers at the decisive point. At Gaza the Turks had approximtely 5,000 men. We had 11,000 sabres, 24,000 rifles, and 170 guns for the capture of this place. But it was naturally strong and had been artificially strengthened. Ali Muntar, the key to the position, was a conical hill on the south-east of the city dominating the town and the country in its vicinity. From this hill due west ran lines of trenches towards Shaikh Hasan. From these trenches there was an excellent field of fire against an enemy advancing from the south. Samson Ridge, a prominent feature, south-west of Gaza, was not fortified, but there were heavy sand-dunes leading up to it from the south. Round Gaza was a series of cultivated plots enclosed by thick, high cactus hedges. The country was all in favour of the defender, and added to the difficulties of a Commander who had to gain his objective in one day. The object was to capture and occupy Gaza before relief could arrive. Supplies admitted of only one day being devoted to the operation. If Gaza was not captured within the time limit the troops would have to withdraw in order to be within reach of food and ammunition. In Gaza there was sufficient water, and it would be possible to draw supplies brought to its vicinity by sea.

Another difficulty which militated against success in the limited time available for achieving it was the weather. Two precious hours of daylight were lost owing to a dense sea fog rolling over the country as soon as the sun rose. The plan for the capture of Gaza was sound, and was to be carried out on the same principle for the capture of Rafa and Magdhaba, only with larger forces, whose movements would take longer to co-ordinate for any operation which had to be undertaken in the course of one day. There was considerable delay in getting into communication with the troops of the 54th Division attached to the 53rd Division for the main attack, and valuable time was lost. This 161st Brigade, with an artillery brigade, did not reach its position from which to attack Green Hill in conjunction with the 158th and 160th Brigades until 1500 hours. This hill was not captured till 1730 hours. It was not till sunset at 1800 hours that the whole Turkish position along the Ali Muntar Ridge was in our hands. Earlier in the day, also, time had been lost. In accordance with the original plan of operations, the cavalry crossed the Wadi Ghazze at Sheikh Nebhan and advanced to Mendur.

One division advanced towards Huj to hold off the enemy's reinforcements, while the other division advanced to Beit Durdis, and later drove the enemy into the northern outskirts of Gaza. The Camel Corps Brigade from the Reserve at Mendur helped in the work of holding off the enemy from the east. The 53rd Division crossed the Wadi Ghazze behind the cavalry, and two brigades of the 54th Division followed behind the 53rd Division and took up a position in support on Mansura and Shaikh Abbas. The infantry brigades reached their first assembly positions on the Mansura Hills by 0830 hours, within 5,000 yards of their objective. Their attack, however, was not launched until nearly midday. Then the 158th Brigade had to advance over the open plain against the enemy in their strong position on Ali Muntar, and the 160th Brigade had to advance up the exposed Sire Ridge.

There was a check close to the Ali Muntar Hill, and the reserve 159th Brigade had to be brought up on the right of the 158th Brigade. The result of all these delays was that we gained our victory just as the sun was beginning to set, and just as the Turkish reinforcements were converging on Gaza from Deir Sineid, from Hureira and from Huj. Gaza, our objective, had not been captured; there was a gap of over two miles between the 54th Division on Shaikh Abbas and the 53rd Division on Ali Muntar. Many of the horses of the mounted divisions had been without water for thirty-six hours, and this water had to be obtained either in Gaza or back in the Wadi Ghazze. It was necessary now to make the decision as to whether the troops should push on to gain the objective or whether to retire. The final situation is summed up in despatches by the Commander-in-Chief : — " If it had now been practicable for the G.O.C. Eastern Force to advance with his three infantry divisions and two cavalry divisions, I have no doubt Gaza could have been taken and the Turks forced to retire, but the reorganization of the force for a deliberate attack would have taken a considerable time; the horses of the cavalry were very fatigued, and the distance of our railhead from the front line put the immediate maintenance of such a force with supplies, water, and ammunition entirely out of the question. The only alternative, therefore, was to retire the infantry."

(8) At the second battle of Gaza on April 19th, 1919, the difficulty of maintaining the objective was considerable at 1500 hours, when the attack had been brought to a standstill. At that time the enemy's strong reserves were still intact, we had not gained Ali Muntar, the key to the position, and the enemy's forward trenches had not been captured. The

Imperial Mounted Division in their dismounted attack on the Atawineh lines had made little progress, though the Birket es Sana spur had been seized and the Imperial Camel Corps Brigade had captured one line of the Khirbet Sihan trenches. The 52nd Division had not been able to advance beyond Lees Hill, Samson Ridge was the limit of the advance of the 53rd Division, the 54th Division had been checked when their left flank was severely enfiladed from Ali Muntar, and they had been heavily counter-attacked when their flank was exposed in its advance beyond the right flank of the centre division. Our tanks were out of action. The 74th Division and two brigades of the 52nd Division had not been seriously engaged. Our casualties, however, had been 6,400, and further attacks against the key of the enemy's position could only be made on a narrow front via Outpost and Middlesex Hills dominated by Ali Muntar. The Commander of the force decided to consolidate our position from Mendur, running in a northerly direction to Shaikh Abbas, where a salient angle was created, as our line ran in a north-westerly direction through Mansura, Samson Ridge to the sea at Shaikh Ajlin. There was no longer any reasonable prospect of success if offensive operations were continued with the available forces in view of the great strength of the enemy's positions.

III. ECONOMY of FORCE which compels a dissipation of the enemy's forces is the constant aim of every commander.

(1) The Turks in the original distribution of the troops between Beersheba and the sea did not economize their forces. They had a wide front with comparatively little depth. The bulk of their twenty-five battalions was split up in holding the series of localities, viz., Beersheba, Irgeig, Hureira, Sheria, Baha, Atawineh, Sihan, Gaza—on a thirty-mile front.

Some dispersion was necessary for them in order to force us to make a wide detour to turn their flank at Beersheba to get the water there, and to cover their railheads north of Gaza, at Hureira and at Beersheba.

With their excellent means of intercommunication they could, however, have held these localities more lightly and in depth at Beersheba as well as at Gaza. Then they might have had a larger central reserve.

(2) The Turks held Amman, Salt, and Shunet Nimrin with weak detachments. These caused no dissipation of force on the part of the E.E.F., while the Turks were liable to be cut off by raiding parties.

In the first raid, carried out from March 21st to April 2nd, 1918, by the 60th Division, Australian and New Zealand Division, Camel Corps Brigade, Mountain Artillery Brigade, Light A.C. Brigade, and heavy battery, the Turks suffered heavily.

There were 953 prisoners taken, five miles of railway south of
Amman were blown up, and a bridge north of the town was
destroyed.

The result of this raid was a further dissipation of the
Turkish forces by increasing the strength of the detachment
there to 8,000 and at Shunet Nimrin to 5,000.

In the next raid carried out by the 60th Division, less one
brigade, the D.M.C., less one division, and the Imperial Service Brigades, between April 30th and May 4th, the Turks
suffered considerably.   Salt was occupied by the mounted
troops, where machine guns and motor vehicles were destroyed.
At Shunet Nimrin 942 prisoners were captured.

The Turks, owing to this raid, further dissipated their forces
by adding to the garrisons at their posts east of the River
Jordan.

(3) After the capture of Junction Station on November
14th the Turks dissipated their forces by sending the
Eighth Army (XX and XXI Corps) up the Sharon Plain
to the vicinity of the River Auja north of Jaffa; the
Seventh Army (III and XV Corps) and 3rd Cavalry Division were sent into the hills to cover Jerusalem.   This did not
cause a similar dissipation in the E.E.F., which thereby
economized its forces by sending the Yeomanry, 52nd and 75th
Divisions and 10th Australian Light Horse Regiment to attack
their eastern force, while the Commander-in-Chief contained
the Eighth Army with the Australian and New Zealand Division, supported later by the 54th Division, which moved up
from Gaza.

(4) The Turks dissipated their forces in the initial operations, which resulted in their withdrawal from the Gaza-
Beersheba line.

Owing to our bombardment of Gaza, starting on October
26th, the Turks thought that our main attack would be in the
coastal area, and sent there a division from their General
Reserves.   They used eight battalions and three cavalry
regiments on their left flank in an attempt to retake Beersheba
from the north.

This use of their reserves helped us in our attacks on
November 6th, when with few casualties our 10th, 60th, and
74th Divisions were able to capture their strongly fortified
positions on a front of seven miles from the Wadi Sharia to
the east of the Hureira Redoubt, after an advance of nine
miles.

IV. CO-OPERATION of all arms at the decisive time and place
    gives the best possible results in war.

(1) The basis of the whole plan by which the Turks were

signally defeated in the final phases of the operation which started on September 19th, 1918, was co-operation.

The available 383 guns in the coastal area were to bombard the enemy's defences from the railway to the sea at 0430 hours on September 19th, when the R.A.F. bombed the H.Q. of the Seventh and Eighth Turkish Armies respectively at Nablus and Tulkeram, and two torpedo-boat destroyers brought fire to bear on the Turks' main line of retreat by the coast.

Then the infantry overran the Turks' positions. The 60th Division with their left flank on the coast advanced across the Nahr Falik to Tulkeram, the 75th Division also crossed the enemy's main defences, 14,000 yards long, 3,000 yards deep, from Bir Adas to the sea, and passing through their second position at Tireh reached Tayibieh, south of Tulkeram.

The 3rd and 54th Divisions continued this line through Felamieh-Jiyus-Kefr Thilth-Bidieh.

The 75th Division remained in Corps Reserve at Tireh. Owing to the rapid success in breaking through the enemy's positions, the cavalry, as soon as the Turks were cleared from the Nahr Falik area, pursued up the coast to Jelameh and Hudeira. Eighteen miles were covered before midday by the leading cavalry.

Within thirty-six hours the 4th Cavalry Division had ridden eighty miles to reach Beisan. Two hours later the 5th Cavalry Division was at Afule and the Australian Mounted Division was at Jenin.

Thus the main lines of retreat to the north for the Seventh and Eighth Armies were blocked. The R.A.F. and Australian Flying Corps co-operated by bombing the Turks' guns and vehicles which blocked the narrow valley road from Tulkeram to Messudie and Nablus. The Arab mobile column co-operated by attacking at the same time in the vicinity of Deraa, in order to keep available Turkish reserves from the coastal area.

Thus all arms contributed to the total defeat of the Seventh and Eighth Armies, which by September 24th had ceased to exist.

(2) There was excellent co-operation between the 53rd, 60th, and Australian and New Zealand Division in the operations between February 19th and 21st, 1918, in the Jordan Valley, when Jericho was entered and our right flank was made more secure than formerly.

The 53rd and 60th Divisions were to advance in a north-easterly and easterly direction respectively, while the Australian and New Zealand Division co-operated by protecting the southern flank of the 60th Division and by cutting off the Turks retiring from Jericho.

In spite of great difficulties of country, for instance, in

places it was only possible to move in single file under the enemy's machine-gun and artillery fire; and although considerable enemy opposition was encountered by the 53rd Division at Ramman and by the 60th Division at Ras Tawil, yet by 0800 hours on February 21st a New Zealand Mounted Brigade and a battalion of the 60th Division reached Neby Musa, fourteen miles east of Jerusalem, and the Australian Mounted Brigade reached Jericho at 0820 hours.

Finally, a line was occupied by the 60th Division through Jebel Ekteif-Talaat Dumm and Ras Tawil, one regiment being left to patrol the valley.

(3) There was a lack of co-operation between the Seventh and Eighth Turkish Armies after our capture of Junction Station on November 14th.

This was due to our rapid advance, which allowed the Turks no time in which to reorganize their lines of communication and enable them to make one line supply their two armies. Therefore, when the E.E.F. was between their armies at Junction Station, the nearest line by which they could co-operate was the Nablus-Tulkeram road.

The rapidity of our advance is clearly brought out in despatches—

" In fifteen days our force had advanced sixty miles on its right and about forty on its left. It had driven a Turkish Army of nine Infantry and one Cavalry Division out of a position in which it had been entrenched for six months, and had pursued it, giving battle whenever it attempted to stand, and inflicting on it losses amounting probably to nearly two-thirds of the enemy's original effectives."

The Commander-in-Chief was thus able to concentrate his efforts against Jerusalem, while forming a defensive left flank across the Sharon Plain north of Jaffa.

(4) The plan of the attack on the Turks' main position covering Sheria and Hureira was based on the co-operation of all arms. This attack on November 6th against the Kauwukah works was to be carried out by the infantry of the 74th, 60th, and 10th Divisions, supported by artillery.

The 74th Division on the north by 1315 hours had captured two lines 3,000 yards apart and then advanced up to the Wadi Sharia, after capturing several detached works along the railway. The artillery of the 60th and 10th Divisions was able to bombard and cut the wire of the main Kauwukah position farther south, so that the infantry could assault at 1230 hours.

Owing to the artillery co-operation these divisions were successful in capturing the Rushdi system between Sheria and Hureira, and by 1700 hours seized Sheria railway station and reached a position close to the Hureira Redoubt.

The cavalry were in readiness to push through the gap in the Turkish positions made by the infantry and artillery. There was further co-operation by the XXI Corps.

To pin the Turks to their ground at Gaza and to prevent them from reinforcing their troops still in the Hureira Redoubt and at Sheria, the XXI Corps made a series of attacks during the night. Outpost and Middlesex Hills were captured at 2330 hours and Belah Trench-Turtle Hill at 0500 hours.

The enemy in the coastal area was then pursued to the Wadi Hesi by the 52nd Division, less one brigade, west of Gaza, and by the Imperial Service Cavalry Brigade east of this town.

Early on November 7th Hureira, Tepe Redoubt, and Sheria were captured.

The 60th Division then took Tel el Sharia, and advanced a mile beyond it. Cavalry now were able to push through in pursuit towards Jemmaneh and Huj.

(5) For the capture of Beersheba the 74th and 60th Divisions, whose eastern flank was covered by the 7th Mounted Brigade and their western flank by the 53rd Division and the Camel Corps Brigade, were to capture the enemy's position on the east between the Khalasa-Beersheba road to the west at Wadi Saba, a tributary of the Wadi Ghazze.

The artillery, with 100 field guns and howitzers and 20 heavy guns, started by bombarding the enemy's positions and guns on a front of 4,500 yards, so that by 0830 hours the 181st Brigade was able to start the assault on Hill 1070.

Three thousand five hundred yards south-west of Beersheba is Hill 1070, which had to be captured, so that the artillery could then cut the wire in front of the Beersheba defences and enable the further advance to be continued.

When the artillery had done this, the infantry assaulted at 1215 hours, and by 1930 hours the enemy's positions north of the Khalasa-Beersheba road and Wadi Saba were captured. The cavalry co-operated during these attacks by attacking Beersheba at dawn from the hills five miles east of the town, and finally by charging into the town at 1900 hours after occupying Bir Sakaty and Tel Saba.

The result of this co-operation of all arms was that the left flank of the main Turkish position was now exposed to an attack from the east.

V. CONCENTRATION of superior force at the decisive time and place is essential for success in battle.

(1) Before the final offensive operations in this campaign began, the Commander-in-Chief concentrated superior numbers at the decisive point. He decided to break through the

enemy's positions in the coastal area. He therefore concentrated 35,000 rifles, 383 guns, and 9,000 cavalry of the XXI Corps, 60th Division, and D.M.C. (less one division) to break through the Turkish defences 3,000 yards deep between Bir Adas and the sea, as well as their second line from Tireh to the mouth of the Nahr Falik.

These positions were held by the Turks with 1,200 cavalry, 8,000 rifles, and 130 guns. This concentration was complete by the evening of September 18th. The result was " that in thirty-six hours, between 0430 hours on the 19th September and 1700 hours on the 20th September, the greater part of the Eighth Army had been overwhelmed, and the Seventh Army was in full retreat." (Despatches.)

(2) The final attack on Jerusalem was made by the concentration of all available forces.

The 53rd Division advanced from the south; the 60th and 74th Divisions with the 10th A.L.H. Regiment between the 53rd and 60th Divisions advanced from the west and northwest of Jerusalem. There was a general assault at dawn on December 8th.

In spite of the difficulties in climbing the rocky hills, in maintaining intercommunication, and in obtaining artillery support, positions were gained one and a quarter miles west of Jerusalem, through Lifta, Beit Iksa, and Nebi Samwil.

On the following day we secured strong positions astride the Jerusalem-Nablus road four miles north of Jerusalem and across the road to Jericho.

(3) The main plan for the defeat of the Turks holding the Gaza-Beersheba line was to concentrate superior forces against the east of their principal defences at Hureira and Sheria.

Three infantry divisions were to capture these defences, and then the mounted troops by advancing to the coast were to cut off the retreating Turks. The first step in this operation would have been to seize Beersheba.

Para. 5 of Force Order No. 54 clearly shows how this concentration was to take place. It is as follows : —

" The XX Corps will move into position during the night 30th/31st October, so as to attack the enemy at Beersheba on the 31st October south of the Wadi with two divisions while covering his flank, and the construction of the railway east of Shellal with one division on the high ground overlooking Wadis el Suffi and Hannafish. The objective of the D.M.C. will be the enemy's defences from the south-east to the north-east of Beersheba, and the town of Beersheba itself."

G

Thus at the time when the XX Corps and two divisions of the D.M.C. attacked on October 31st, we had concentrated a large superiority of force against the enemy's weak 67th and 81st Brigades and the small proportion of sabres detached from their Cavalry Division in the vicinity of Beersheba.

VI. SECURITY.

(1) On September 19th, when the Commander-in-Chief was breaking through the enemy's defences between the railway and the sea with the bulk of his forces, namely, 9,000 sabres, 383 guns, and 35,000 rifles, his centre and right flank were secured by the XX Corps, consisting of the 53rd and 10th Divisions, and General Chaytor's Force, consisting of the Anzac Mounted Division, 20th Indian Infantry Brigade, two battalions of the British West Indies Regiment, two Jewish battalions, and a heavy battery.

In addition, the Arab Mobile Force operated north-west and south-east of Deraa to prevent available Turkish reserves being sent down to Afule and Beisan. Chaytor's Force covered the eastern flank of the E.E.F.

By September 21st the Force reached Khurbet Fusail, west of the Jericho-Beisan road.

On the following day the N.Z.M.R. gained a position dominating the Wadi Farah road leading to the Jisr Damie bridge, thus blocking the final means of escape of the Turkish Seventh and Eighth Armies.

Chaytor's Force then occupied Amman on the 25th, where they remained in order to intercept Turks retreating from the Hedjaz. The XX Corps carried out their mission by assaulting the enemy on either side of the Jerusalem-Nablus road.

By midday on September 20th the 10th Division had advanced seven miles in difficult hilly country against strong opposition, especially at Furkah. They kept touch with the French contingent on their left and gained positions near Iskaka and at Selfit. The 53rd Division continued the line through Malik and Khan Jibeit and maintained touch with the 10th Division and with Chaytor's Force.

(2) In the raid against the Hedjaz railway, starting on March 26th, 1918, our attacks on Amman on the 28th and during the night 28th/29th March did not enable our two mounted, one infantry and one camel brigade to capture this village. At Salt there was one cavalry and one infantry brigade. There was one infantry brigade between Shunet Nimrin and Ghoraniyeh.

Enemy reinforcements on the 29th assaulted the left flank of our positions at Amman, and we had no fresh troops at hand to deal with them, as a Turkish cavalry division and two brigades of infantry on the morning of this day attacked our troops at Salt.

The floods had swept away all but one of our bridges over the River Jordan. The approaches to the other bridges were under water, as the River Jordan in the course of the morning rose nine feet. The troops were very tired and there was continuous rain.

Our attacks against Hill 3039 south-west of Amman Station were commenced at 0200 hours on September 30th, with one battalion of the Camel Corps Brigade and the New Zealand Mounted Brigade, while the rest of the Camel Corps Brigade advanced south of the Salt-Amman road, and the infantry brigade advanced north of it for the attack on the town. The two A.L.H. Brigades operated farther north.

Some progress was made towards the town and the southern slopes of Hill 3039. Enemy reinforcements, however, pressed throughout the day against our left flank, and we were unable to gain the hill or the citadel of Amman.

The Turks were also attacking at Salt and against our advanced posts east of the River Jordan south of Jisr Damie Bridge. Thus all our available reserves were engaged in fighting. Therefore in the interests of the security of the raiding force it was necessary to withdraw.

The method of securing the withdrawal was for the troops engaged at Hill 3039 to hold a chain of posts on the west bank of the Wadi Amman to cover the infantry and Camel Corps battalions, who withdrew via Ain Sir to avoid the Turks engaged at Salt. The covering force, after the infantry had passed Ain Sir, withdrew fighting a rearguard action.

Our positions had to be held at Salt all through the 31st, as the enemy were in close contact. At 2300 hours they retreated, and our infantry brigade withdrew, covered by the 1st A.L.H. Brigade. The whole force was safely back west of the River Jordan, leaving only a bridgehead on the left bank.

(3) For the security of our right flank, and to be able to raid the enemy's communications, it was necessary after the capture of Jerusalem to capture Jericho, so that it could be used as a base for operations against Amman, and to occupy the lower Jordan Valley, and to control the river crossings in this area in order later to extend the front.

The Commander-in-Chief, therefore, decided to send the 60th Division and the Australian and New Zealand Division in an easterly and north-easterly direction. The 60th Division on February 19th, after capturing El Muntar, Arak Ibrahim, and Ras Tawil, had their left flank secured by the 53rd Division and their right flank secured by the mounted troops.

On the following day, when the 60th Division advanced to Jebel Ekteif, the 53rd Division was in a commanding position from which to secure their left flank.

The Australian and New Zealand troops secured their right flank by crossing the Wadi Kumran south of El Muntar and gaining a position in the Wadi Jofet Zeben. By the 22nd the line through Elteif Hill-Talaat Dumm—Ras Tawil was occupied and secured by outposts east of these places, and by a regiment of the Australian and New Zealand Division patrolling the valley.

It was, however, further necessary to enlarge our position in order to obtain a suitable base from which to start operations against the Hedjaz railway. There is a prominent ridge ten miles long and two miles broad running from Musalabeh on the north to Sultan Hill on the south. If this ridge was in our possession we should command the approaches to the Jordan throughout its length and prevent the Turks transferring troops from the west to the east bank in this area, and we should prevent the enemy from using the Ghoraniyeh bridge.

To secure the left flank of the 60th Division, who captured Beigudat and Tellul Hill on March 10th, and the 53rd Division, who advanced on to Kefr Malik and took Asur Hill, the 74th Division on the left of the 53rd Division advanced across the Wadi Nimrin when the 10th Division crossed the Wadi Jib and gained, in both cases, the heights on the far side by March 11th.

On the following day the operations were completed. The left flank of the XX Corps, which was established on Felah Hill and on hills commanding Sinjil, was secured by the 75th Division advancing from the Wadi Ballut Ridge to Benat Bumy, and with the 54th Division captured a series of positions from Mezeireh to Mejdal Yaha, Ras el Ain, and El Mirr.

(4) Between December 26th and 30th, 1917, the Turks attempted to retake Jerusalem. Their attacks started against the 60th Division at 2330 hours on the 26th, and again at 0130 hours on the 27th, the 19th, 24th, and 53rd Divisions of the enemy engaged this division on its whole front on both sides of the Jerusalem-Nablus road, their main attacks being delivered against Ful Hill. They were successfully repulsed.

Night attacks were also commenced by their 26th and 27th Divisions against the 53rd Division east of Jerusalem. These were also successfully dealt with, and where the enemy did penetrate into our positions they were driven out by reserves. The front of the 53rd and 60th Divisions was finally secured by counter-attacks delivered by 74th and

10th Divisions against the Turks' right flank from Zeitoun Ridge towards Ram Allah.

In spite of opposition, these divisions advanced 4,000 yards on a six-mile front.   The 74th Division gained a position opposite Beitunia and the 10th Division occupied Deir Abzia. They drew off the enemy's reserves from their attacks on the 53rd and 60th Divisions, and stopped their attempts to regain Jerusalem.

"As a result of three days' fighting, not only did the enemy's attempt to recapture Jerusalem fail, but by the end of the third day he found himself seven miles farther from Jerusalem than when his attack started."   (Despatches.)

(5) Throughout the campaign the Air Force added to the security of the E.E.F. by their reports as to the enemy's positions and movements.

After Junction Station had been taken on November 14th they reported that the Turks would probably reorganize their forces on the Tulkeram-Nablus road.

They were able to report on November 11th that the Turks on the general line covering Beit Jibrin, Summeil, Tunnus, and Burka on the Nahr Sukereir numbered about 15,000, and that they intended to try to save Junction Station from being captured.

They also reported on the preparations made by the Turks to strengthen their positions between Gaza and Beersheba. The following is a quotation from Despatches to illustrate the value of the reports obtained at G.H.Q. : —

" It was evident from the arrival of reinforcements and the construction of railway extensions from Tine on the Ramleh-Beersheba railway to Deir Sineid and Beit Hanun, and from reports of the transport of large supplies of ammunition and other stores to the Palestine front, that the enemy was determined to make every effort to maintain his position on the Gaza-Beersheba line."

(6) Before the attack against the Turks holding this line, the concentration of the XX Corps and D.M.C. in their positions from which to operate on October 31st was carefully secured by an outpost of the two D.M.C. and 53rd Divisions from Esani-Buggar to the east of Goleib.

The Australian and New Zealand Divisions were at Khalasa and Asluj, which they reached by night marches.

The 60th and 74th Divisions reached their positions between Esani and Gamli also by night marches.   They remained concealed in wadis during the day.   All these positions, within striking distance of the enemy's entrenchments, were reached without attracting their attention.

(7) The Turks secured their front of thirty miles from Beersheba to the sea by a series of works all strongly entrenched and wired. The main positions were at Gaza, Sihan, Atawineh, Baha, Sheria, Hureira, Irgeig, and Beersheba. The weakness of these positions was that north-west of Beersheba there was a gap of four and a half miles.

The other works were mutually supporting about a mile apart. However, they added to their security by their communications, which were improved, so that any threatened point in their line could be quickly reinforced.

VII. OFFENSIVE ACTION is necessary to obtain victory.

(1) The Turks relied on holding strong positions. They used up the bulk of their troops in this way, keeping a very small proportion as a general reserve for offensive action. On no occasion did their whole force take the offensive.

The result was that they suffered a series of defeats in the loss of these positions, and finally the E.E.F. gained a victory by the continued offensive which ended in the destruction of the Seventh, Eighth and Fourth Armies and our occupation of Aleppo.

(2) Offensive action characterized the whole of our final operations.

The plan aimed at the destruction of all the Turkish forces west of the River Jordan on a front of forty-five miles distributed to a depth of twelve miles. Once this has been accomplished, the offensive could be continued east of the River Jordan, so that touch could be obtained with the Arab Army, and so that the communications with Constantinople could be cut and the isolated Fourth Army east of the River Jordan could be dealt with decisively.

For the offensive operations to be decisive west of the River Jordan the infantry, supported by artillery and the fire of two ships, were to break through in the coastal area as early as possible.

Our great superiority of force was to be used at the one place where the mounted troops could be used to push the offensive operations to the utmost limit. The initial success was complete. After a fifteen-minute bombardment in the coastal plain, the infantry attacked from the Deir el Kussis Ridge to the sea.

The French contingent, 54th, 3rd, 75th, 7th and 60th Divisions, quickly overcame the enemy's resistance at Wadi Karah, Bir Adas, Tabsor, and along the coast. Their second line of defence was then overwhelmed, and when the 60th Division had cleared the Nahr Falik and the marshes in the vicinity of the enemy, and advanced north-east on

Tulkeram, the mounted troops were able to begin their advance.

At 0630 hours the 5th Cavalry Division began to advance in pursuit of the defeated Eighth Turkish Army. The R.A.F. now co-operated in low-flying action with bombs and machine guns against the enemy, blocked on the Nablus-Jerusalem road.

The offensive was continued towards Nablus by the 53rd and 10th Divisions against the Seventh Army as soon as the Eighth Army had been driven back. By midday on the following day (20th) the 10th Division had advanced seven miles. Messudie, which was on their line of retreat over the mountains of Samaria to Jenin, was in our possession.

On the following day Nablus was secured and the Seventh Army was in retreat.

The offensive was continued by the infantry, so that all organized resistance by the enemy rearguards ceased on September 21st. The cavalry, who were astride their main lines of retreat by September 23rd at Afule and Beisan, with Haifa, Acre, Tiberias, and Mejamie Bridge in their possession, were able to deal effectively with the retreating Seventh and Eighth Armies, which, as armies, ceased to exist by September 24th.

The offensive was then continued by Chaytor's Force advancing on Amman against the Fourth Army, whose position east of the River Jordan was no longer tenable.

This place was captured on September 25th. The garrison of Maan, 5,000 strong, surrendered to Chaytor's Force on September 28th.

The D.M.C. now continued the offensive by advancing on Damascus to cut off the remainder of the Turkish forces and to occupy this city. This was carried out by October 1st. The Fourth Army had then ceased to exist.

The XXI Corps advanced up the coast from Haifa to Beirut, which was reached by October 8th. Here a post for supplies was opened on October 10th.

Success was now further exploited. The 5th Cavalry Division after a march of eighty miles occupied Homs on October 15th, Tripoli on the coast having been occupied by the XXI Corps Cavalry Regiment and armoured cars two days earlier.

A bold advance was now possible owing to the demoralization of the enemy after our continuous pressure since September 19th, so that one cavalry division was now able to deal with greatly superior numbers of the enemy in the final phase.

These final offensive operations resulted in the occupation of Aleppo by the 5th Cavalry Division astride the enemy's communications with Aleppo on October 26th. On October 31st the Armistice with Turkey was completed.

In this victory 75,000 prisoners were captured.

(3) The limit to which offensive operations can be pressed must always be carefully considered. There should be a superiority or advantage over his opponent, which will enable a commander to take the offensive, but there will also be a time when these advantages, owing to casualties or necessity for the reorganization of communications, will decrease.

After our successful operations which led to the capture of Jerusalem and positions securing the port of Jaffa, it was not possible to continue our offensive immediately. It was necessary to improve communications, as the Turks had destroyed the bridges of the Jaffa-Jerusalem railway in the hills. Supply arrangements in the hilly country, in which there was only one road suitable for vehicles, required to be reorganized.

The six Turkish divisions were in strong positions astride the roads leading from Jerusalem to Jericho and Nablus. The winter weather would add to the difficulties of hill fighting. Also the E.E.F. had to be reorganized by replacing British cavalry and infantry with Indian cavalry from France and Indian infantry from India.

Therefore, the Commander-in-Chief decided to enlarge and improve his positions by a series of operations which would give sufficient space for manœuvre, and from which the final offensive could be undertaken with the best possible prospects of continuing it until a victory was obtained.

(4) After the defeat of the enemy on the Gaza-Beersheba line the Desert Mounted Corps, supported by the 52nd and 75th Divisions, vigorously continued offensive operations.

G.H.Q. on November 9th, 1917, ordered the D.M.C. at Huj "to press the enemy relentlessly. The XXI Corps was to get troops forward towards Julis-Mejdel. D.M.C. was to press on by the shortest route with all available forces to Tine-Beit Duras." (Despatches.)

The numbers available for these offensive operations against the Eighth Army were limited by the transport available for keeping them supplied beyond railhead. The 10th, 74th and 60th Divisions had to return to Belah and the 54th Division remained at Gaza. However, with the available two infantry and three cavalry divisions so much pressure was maintained against the enemy that by November 12th a further advance of 20 miles had been made up to the Wadi Sukereir.

(5) In the operations for the capture of Beersheba on October 31st the mounted troops successfully reached their allotted positions five miles east of the town at the appointed time. However, they then encountered considerable opposition, as the ground on the east and north-east was flat and was commanded and flanked by Tel Saba.

The hill was strongly held. It was essential to capture it before an advance in force over the plain could be carried out. Only small parties were able to advance during the afternoon across the plain until this hill was captured. Then it was decided to carry out a mounted attack.

This offensive operation was completely successful. The 4th Australian Light Horse Brigade at 1830 hours charged over two lines of enemy trenches outside the town, supported by machine guns and guns. The Turks were surprised and completely defeated. The town was entered at 1900 hours, and 2,000 prisoners and 13 guns were captured.

(6) The actions at Magdhaba, December 23rd, 1916, and at Rafa, January 9th, 1917, were excellent examples of sustained offensive actions.

In these actions the E.E.F. cleared the Turks out of Egypt. The Turks had detachments at these places, as they evidently considered that they were out of reach of our infantry, and that their fortified posts were strong enough to resist attack by mounted troops.

The Anzac Mounted Division and Camel Corps Brigade, however, made a night march of twenty-five miles up the Wadi el Arish, and by daybreak they were within four miles of their objective. The Turks' position was in a circle of five self-contained redoubts on each side of the Wadi el Arish. Two redoubts were east of Magdhaba, and three redoubts were west of the village. Our mounted brigades made converging attacks against all these redoubts, and by midday had practically surrounded the position covered by the fire of three batteries. For eight hours the attacks were pressed with great determination.

Then a combined attack by the 3rd Australian Light Horse Brigade working up the wadi and two companies of the Camel Corps Brigade advancing from the north-west caused No. 1 Redoubt to fall with 3 officers and 92 other ranks. This success caused the whole defence to collapse just as darkness was falling, and just as the Commander's orders for a withdrawal were being circulated. Had these offensive operations not been successful by this time withdrawal would have been necessary, as there was no water available for the horses.

The order to retire was cancelled soon after it was issued, that is, when No. 1 Redoubt fell.

The success of the operation was due to the continuous offensive throughout the hours of daylight. The Turkish Commander, two battalion commanders, 1,280 men and 4 guns were captured at a loss to us of 22 killed and 124 wounded.

(7) The position at Rafa consisted of three strong posts supported on their north side by a central redoubt on a prominent knoll. The approach to these works was over open ground devoid of any cover. The Turks had an excellent field of fire in all directions up to 2,000 yards.

In spite of these advantages for the Turks our attacks were vigorously pressed on January 9th, 1917. On the evening of January 8th the Anzac Mounted Division (less one brigade) the 5th Yeomanry Brigade, the Imperial Camel Corps Brigade, and six motor machine-gun cars left El Arish, and, after a twenty-nine-mile night march, reached the Turkish position at Magruntein, 2,000 yards south-west of Rafa. Offensive operations were at once commenced. The 3rd and 1st Light Horse Brigades attacked the Turkish works facing east. The Camel Corps Brigade attacked the works facing south, the 5th Mounted Brigade attacked the works on the western side, while the New Zealand Mounted Rifles attacked the redoubt from the north. Progress was slow owing to the strength of the enemy's positions. About 1600 hours it was seen that reinforcements, about 500 strong, were arriving from the direction of Khan Yunius and Shellal. At this time the Commander of our force gave the order to withdraw, for as at Gaza on March 26th, 1917, and at Magdhaba on December 23rd, 1916, as evening approached the necessity for watering the horses became imperative. However, in this case the situation was saved by the successful offensive operation carried out by the New Zealand Mounted Rifle Brigade, who made a fine bayonet charge against the redoubt and captured it. Then the Camel Corps Brigade was successful in the attack on their objective. Five officers and 214 other ranks surrendered to them. After this the garrisons of the other two works surrendered before our attacking troops reached their parapets. Complete victory was obtained by the continuous offensive; 1,635 prisoners, 4 guns, and 6 machine guns were captured.

VIII. MOBILITY.

(1) The destruction of the Turkish Seventh and Eighth Armies was largely due to the mobility of the cavalry. The basis of our plan was for the XXI Corps to break through in

the coastal area and to drive the Turks there in an easterly direction towards the hills, so that the cavalry could operate in ground suitable for rapid action. Thirty-six hours after the attack started on September 19th, 1918, the cavalry were astride the enemy's main lines of retreat.

Three hours after they started their pursuit, the leading troops of the D.M.C. had covered eighteen miles. Before dark the 5th Cavalry Division were at Hudeira; the 4th Cavalry Division were on the northern flank of the infantry positions about Jelameh.

By the following evening the 5th Cavalry Division crossed the mountains of Samaria at Shusheh into the Plain of Esdraelon and reached Afule and Nazareth; the 4th Cavalry Division crossed these mountains at Musmus Pass and reached Beisan via Afule by 1630 hours. This division had now marched eighty-five miles in thirty-four hours, and had captured 1,400 prisoners, one regiment being detached to seize the Mejamie Bridge.

Two brigades of the Australian Mounted Division crossed the Musmus Pass, and then advanced from Lejjun in a south-easterly direction to Jenin.

On September 23rd the 13th Cavalry Brigade occupied Acre, and the 15th Cavalry Brigade, after dashing attacks from the east and south, captured Haifa. The Seventh and Eighth Turkish Armies by September 23rd ceased to exist.

The D.M.C. was now ordered to cut off the retreat of those of the Fourth Army who had escaped from Chaytor's Force, and to occupy Damascus. On September 25th the Australian and New Zealand Division captured Amman.

The advance was now made in two columns. The Australian and 5th Cavalry Divisions from Nazareth marched via Tiberias, and crossed the Jordan south of Lake Huleh, after stiff fighting at Benat Yakub Bridge. They reached Kuneitra on the 28th, Sasa on the 29th, and the vicinity of Katana on the 30th. The 4th Cavalry Division crossed the Jordan below Lake Tiberias at Mejamie Bridge.

On September 27th they were at Irbid and Remte; on the 28th at Mezerib; on the 29th in touch with the Arab mobile force at Ezra, approximately 550 miles from their starting-point.

(2) The mobility of an army depends on the supply and transport services.

Before the advance of the E.E.F. on September 19th, all arrangements were made to expedite the rapid continuance of the offensive.

A supply depot was formed behind the area occupied by the 54th Division, where a railhead was established. Water-supply was arranged up to Jiljulie by the evening of the first day of operations, as the necessary pipes had been brought during the previous nights up to our forward positions held by the 54th Division. A light railway was constructed from Jaffa up to our positions on the coast and from Jerusalem to Bireh in the Judean Hills. The Ludd-Jerusalem railway was converted into standard gauge.

The port of Jaffa was improved. At Ludd large supply and ordnance depots were made. Roads were made from Ludd due north to our positions and were connected up with the Jerusalem-Bireh road, which was improved.

(3) The basis of the final plan, by which the Commander-in-Chief was able to defeat the Seventh and Eighth Turkish Armies west of the River Jordan, and to isolate the Fourth Turkish Army and then to deal decisively with it, was for the cavalry to pass through the gap made in the enemy's positions in the coastal area, and to block their line of retreat in the Plain of Megiddo in the Valley of Jezreel by occupying Afule and Beisan.

In order to gain full value from the mobility of cavalry, this gap in the Turkish positions was made in the Plain of Sharon between Jiljulie and the coast by the 54th, 3rd, 75th, 7th, and 60th Divisions pivoting on the French contingent and supported by artillery and naval guns.

The cavalry were thus able to cross the mountains of Samaria by the Musmus Pass and by Shusheh more quickly than the Turks could by the mountains into which they were driven by the XXI Corps.

(4) On April 30th, 1918, the Australian Mounted Division in the raid on Salt reached this village thirteen miles from Shunet Nimrin and captured it one hour and ten minutes after leaving the 60th Division, who were attacking the last-named village. Full value was thus made of the mobility of the cavalry to cause the Turks to split up their forces and to divert them from concentrating at Shunet Nimrin.

(5) Owing to the rapidity with which the D.M.C. had pressed back the Turkish rearguards along the coast north of Gaza, the enemy by November 13th were spread out over a front of twenty miles from Beit Jibrin to Kubeibe, covering Junction Station on their left with positions about five miles from it.

The Turks made their most determined effort to defend Junction Station in their defence of Katrah and Mughar. The 52nd Division was checked in front of this position by 1400 hours.

The 6th Mounted Brigade, with the Berks Horse Artillery Battery and the 17th Machine Gun Squadron, was in a position of readiness in the Wadi Jamus behind the left flank of 135th Brigade, who were in the Wadi Ghor west of Mughar.

At 1500 hours, after a reconnaissance of the ground so as to gain the full co-operation of guns and machine guns in their attack, they issued from the wadi and made full use of their mobility by turning the enemy's right flank, and by gaining a position on the hills north of Katrah village, in spite of the fact that the last 3,000 yards of their advance was under heavy gun and machine-gun fire across open ground commanded from Mughar. They were supported by machine guns from Wadi Ghor and the horse battery firing at a range of 3,200 yards.

Once on these hills north of the village, the enemy's position became untenable, as the 155th Brigade then made a frontal attack, and with the cavalry closing on the Turks' right flank captured 1,100 prisoners, 12 machine guns, and 2 guns.

(6) The basis of the original plan for the capture of the Gaza-Beersheba line was to make full use of our superior mobility. Two mounted divisions were to make the enveloping movement round the east of Beersheba, so that later the whole of the enemy's position could be turned by attacking the Hureira-Sheria defences from the east. The cavalry would then be in country suitable for their employment for cutting off the Turks in the Atawineh and Gaza positions, and for the pursuit of those who escaped.

On the night October 30th/31st two mounted divisions were, respectively, in position at Khalasa and Asluj. In spite of long night marches of twenty-five and thirty miles respectively, they were in their positions by 0800 hours on the following morning, and completely surprised the Turks holding the eastern Beersheba defences.

The 60th and 74th Divisions captured Hill 1070 and the main defences south-west of the town. The final mounted attack at 1830 hours by the 4th A.L.H. Brigade across the open plain from the east of the town made this preliminary operation a complete success.

BIBLIOGRAPHY.

The Military History of the Great War; Military Operations in Egypt and Palestine.

The Official History of Australia in the War. (H. S. Gullett.)

The Revolt in the Desert. (T. E. Lawrence.)

Soldiers and Statesmen. (Sir William Robertson.)

L'attaque du Canal de Suez. (Douin.)

INDEX

95

.

MAP I.

SCALE OF MILES

0   5   10   20

Railways ....
Roads ....
Sweet Water Canal

Port Said

Lake

Menzala

Tina Plain

Pelusium

Lake Bardawil

Turks' advance

3 Bns. Mountain Battery

Abd

Qatiya

28 Jan.'15 Dueidar

Bir el Nus

Qantara

2000 Turks 1st Feb.

Hod el Aras

NORTHERN SECTOR.
1 Sqn. Cav., 2 field
Mountain Bty.
½ Coy. R.E., 5 Bns.
1 Armoured Train
with ½ Coy. Inf.

Ballah Lake

Turks had
10th, 23rd, 25th Divs.
each 4000 strong approx.
and 10,000 camels.
Also 28th Cav. Regt.
9 field batteries.
2 6" Howitzers.

2 Coys.

Ferdan

Ismailia

5000 Turks 1st Feb.

Timsah

2 Coys. 2 m.g.

Kress von Kressenstein's Force
1 Sqn. Cav., 2 Mountain Btys.
3 Battalions left in desert
after 8th Feb. 1915.

Tussum

5000 Turks 1st Feb.

CENTRAL SECTOR.
1 Sqn. Cav.
1 Bde. Fd. Arty.
1 Mountain Bty.
7 Battalions
2 ½ Coys. Camel Cps.
1 m.g. " " sec.

Serapeum 3 Coys.
Deversoir 2 Coys. 2 m.g.

Turks' advance

6 Bns. Heavy Bty. Pontoons

Great Bitter Lake

Moiya Harab

Geneffe

SOUTHERN SECTOR.
1 Squadron
1 Infantry Brigade
1 Field Battery
½ Coy. Roy. Eng.
1 Coy. Camel Corps.

27 Jan.'15
Kubri

Suez

Gebel Murr

Ain Musa

Turks' advance

2 Bns. Mtd. Tps.

GULF OF SUEZ

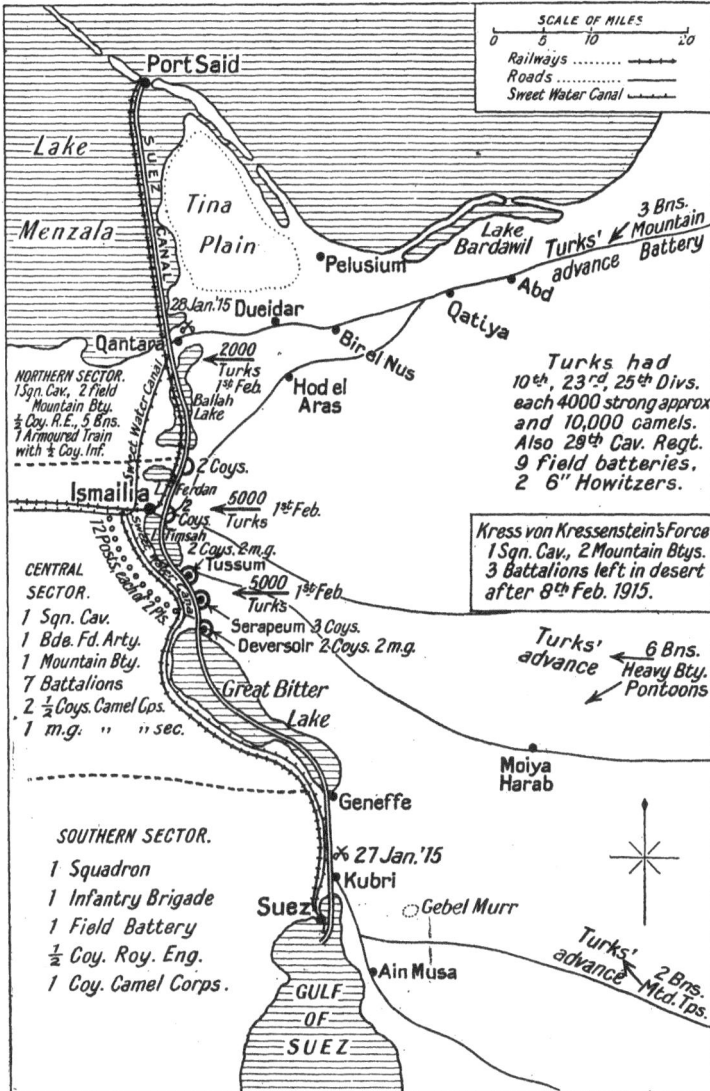

THE ADVANCE BY THE TURKS ON THE SUEZ CANAL, 1915.

MAP 2.

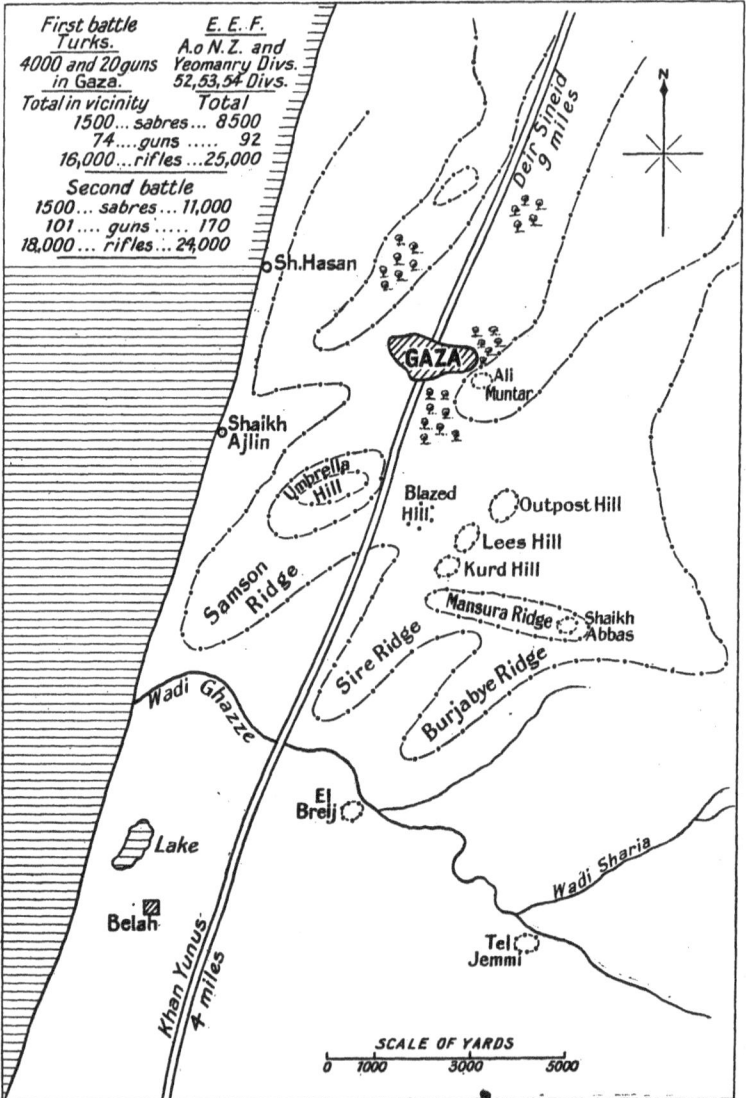

First battle
Turks.
4000 and 20 guns
in Gaza.
Total in vicinity
1500 ... sabres ... 8500
74 .... guns 92
16,000 ... rifles ... 25,000
Second battle
1500 ... sabres ... 11,000
101 .... guns 170
18,000 ... rifles ... 24,000

E. E. F.
A.o N.Z. and
Yeomanry Divs.
52,53,54 Divs.
Total

Sh. Hasan

GAZA

Ali
Muntar

Shaikh
Ajlin

Umbrella
Hill

Blazed
Hill

Outpost Hill

Lees Hill

Kurd Hill

Samson
Ridge

Mansura Ridge

Shaikh
Abbas

Sire Ridge

Burjabye Ridge

Wadi Ghazze

El
Brejj

Lake

Wadi Sharia

Belah

Tel
Jemmi

Khan Yunus
4 miles

Deir Sineid
9 miles

N

SCALE OF YARDS
0   1000   3000   5000

GAZA BATTLEFIELDS, MARCH 26TH AND APRIL 19TH, 1917.

106

MAP 3.

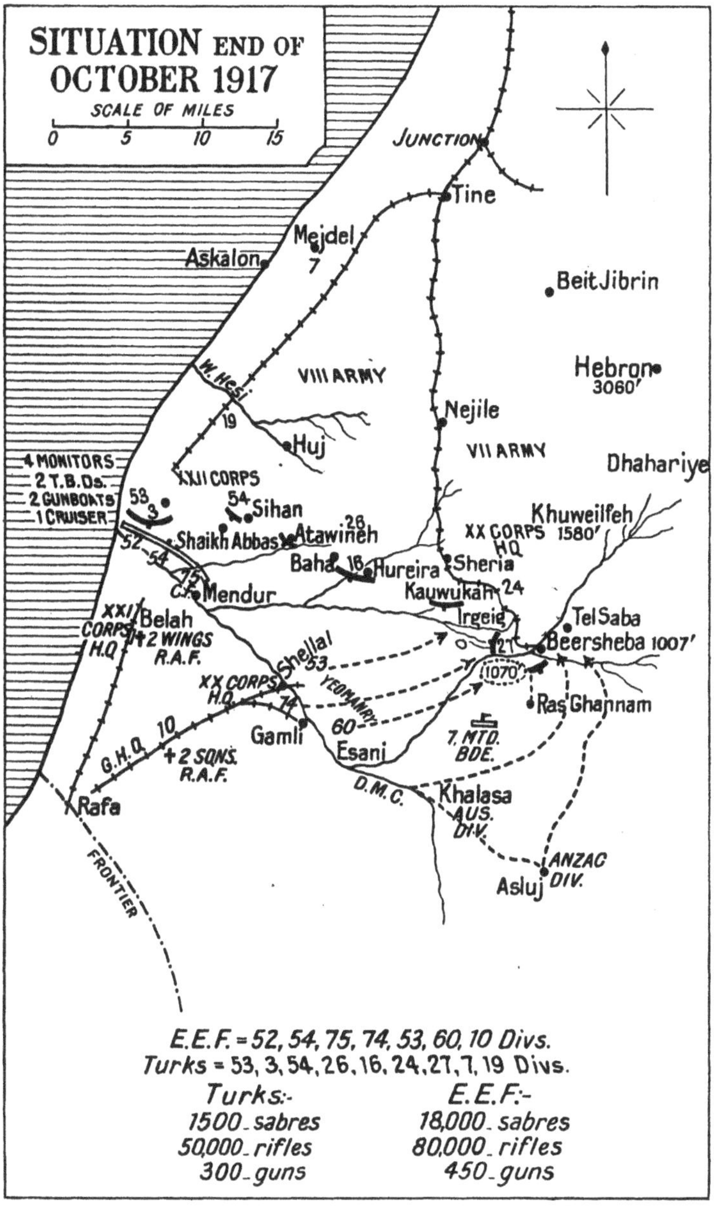

SITUATION END OF
OCTOBER 1917

SCALE OF MILES

0   5   10   15

JUNCTION

Tine

Mejdel
7

Askalon

Beit Jibrin

W. Hesi

VIII ARMY

Hebron
3060'

19

Nejile

Huj

VII ARMY

Dhahariye

4 MONITORS
2 T.B.Ds.
2 GUNBOATS
1 CRUISER

XXII CORPS

58
3

54 Sihan

Khuweilfeh
1580'

52
54

Shaikh Abbas

Atawineh
26

XX CORPS
H.Q

17

Baha 16

Hureira

Sheria

C.F. Mendur

Kauwukah 24

XXI
CORPS
H.Q.

Belah
+2 WINGS
R.A.F.

Shellal
53

Irgeia

Tel Saba

21

1070

Beersheba 1007'

XX CORPS
H.Q.

14

YEOMANRY

Ras Ghannam

G.H.Q

10

Gamli

60

7. MTD.
BDE.

+2 SQNS.
R.A.F.

Esani

Rafa

D.M.C.

Khalasa
AUS.
DIV.

FRONTIER

Asluj

ANZAC
DIV.

E.E.F. = 52, 54, 75, 74, 53, 60, 10 Divs.
Turks = 53, 3, 54, 26, 16, 24, 27, 7, 19 Divs.

Turks:-
1500_sabres
50,000_rifles
300_guns

E.E.F:-
18,000_sabres
80,000_rifles
450_guns

MAP 4

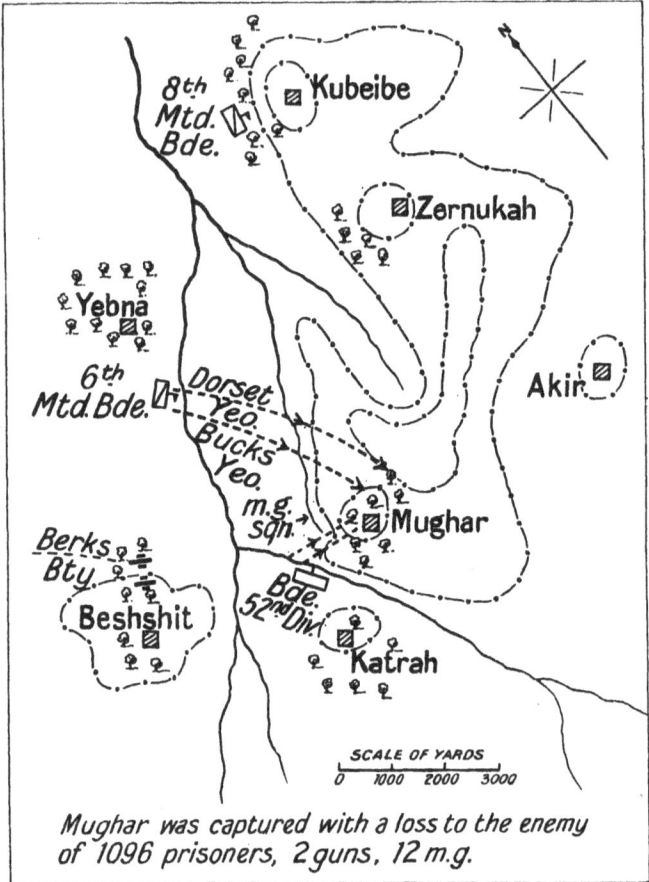

Mughar was captured with a loss to the enemy
of 1096 prisoners, 2 guns, 12 m.g.

BATTLE AT MUGHAR, NOVEMBER 13TH, 1917.

Map 5.

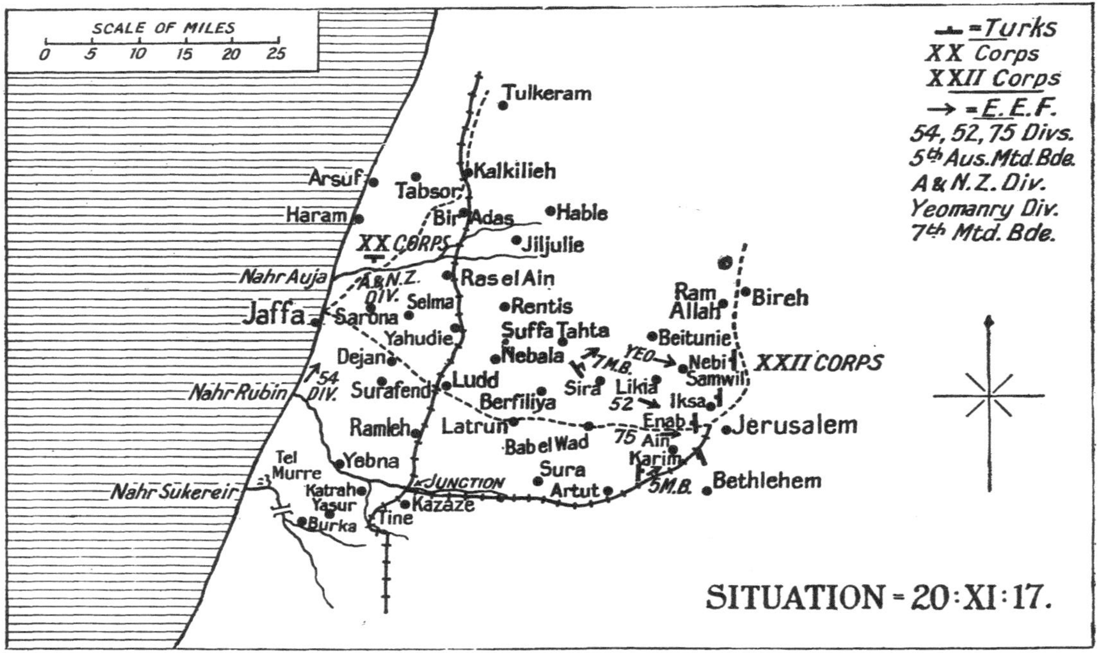
SCALE OF MILES
0 5 10 15 20 25
= Turks
XX Corps
XXII Corps
→ = E.E.F.
54, 52, 75 Divs.
5th Aus. Mtd. Bde.
A & N.Z. Div.
Yeomanry Div.
7th Mtd. Bde.
Tulkeram
Arsuf
Tabsor
Kalkilieh
Haram
Bir Adas
Hable
XX CORPS
Jiljulie
Nahr Auja
A & N.Z. DIV.
Ras el Ain
Jaffa
Sarona
Selma
Rentis
Yahudie
Suffa Tahta
Ram Allah
Bireh
Beitunie
Dejan
Nebala
7 M.B.
YEO
Nebi Samwil
XXII CORPS
54 DIV.
Nahr Rubin
Surafend
Ludd
Berfiliya
Sira
Likia
52
Iksa
Ramleh
Latrun
Enab
Jerusalem
75
Ain Karim
Bab el Wad
Tel Murre
Yebna
Sura
5 M.B.
Bethlehem
JUNCTION
Artut
Nahr Sukereir
Katrah
Yasur
Kazaze
Burka
Tine
SITUATION = 20 : XI : 17.

MAP 6.

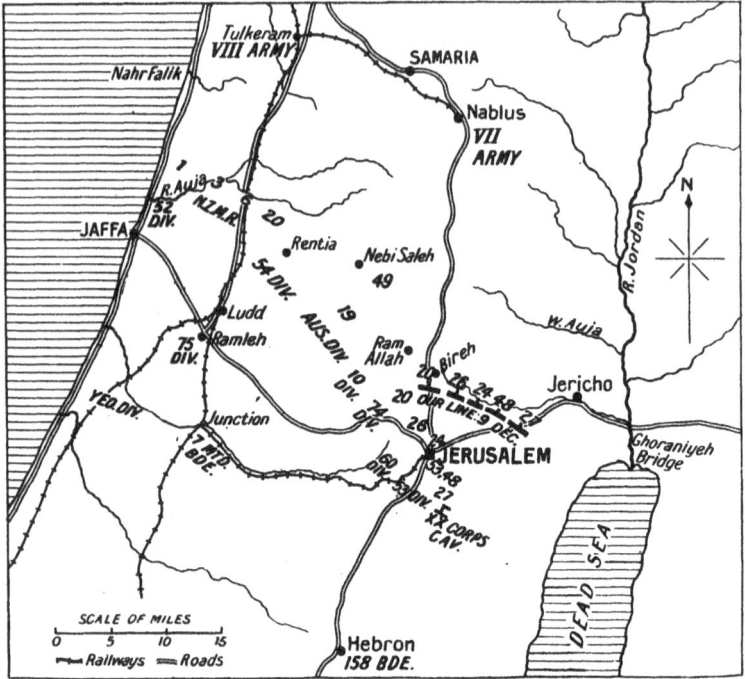

FINAL OPERATIONS FOR CAPTURE OF JERUSALEM, DECEMBER 9TH, 1917.

MAP 7.

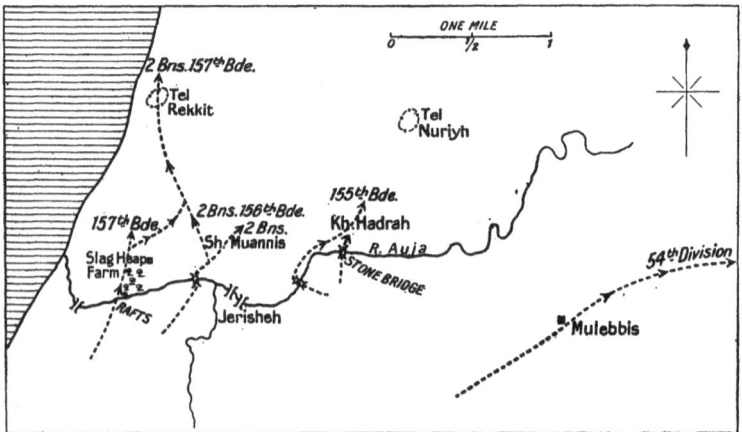

CROSSING OF THE RIVER AUJA, NIGHT OF DECEMBER 20TH/21ST, 1917.

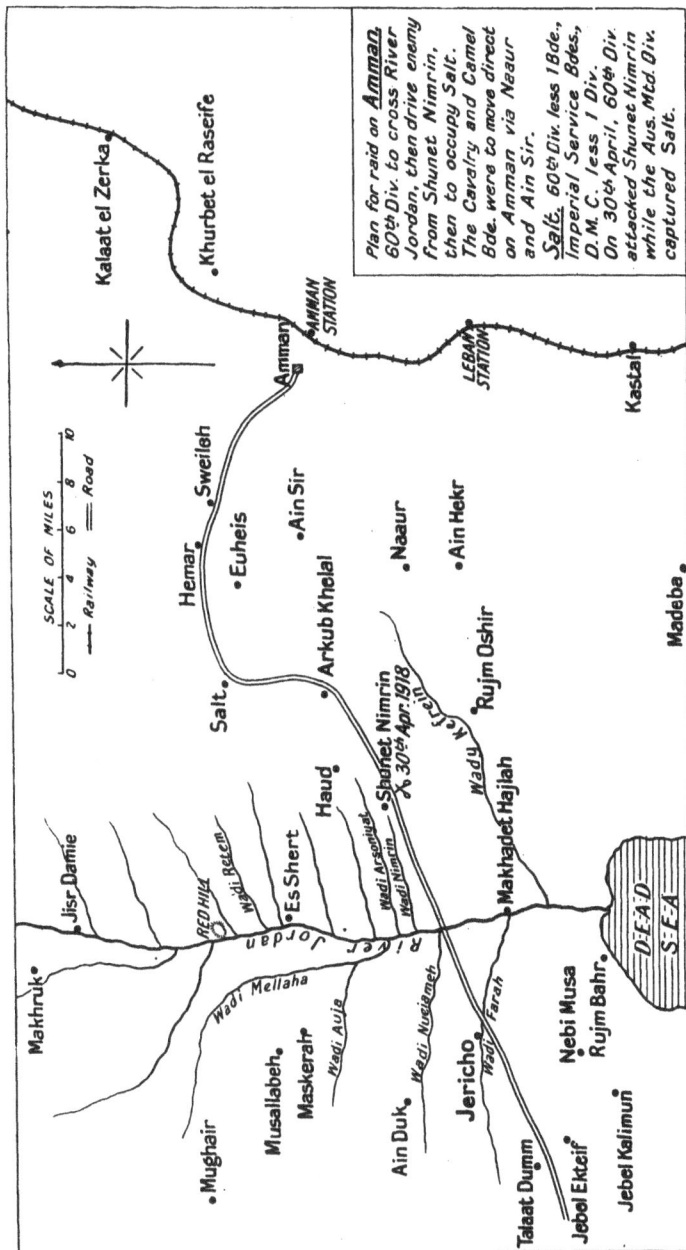

MAP 8.

Plan for raid on **Amman.**
60th Div. to cross River
Jordan, then drive enemy
from Shunet Nimrin,
then to occupy Salt.
The Cavalry and Camel
Bde. were to move direct
on Amman via Naaur
and Ain Sir.

Salt. 60th Div. less 1 Bde.,
Imperial Service Bdes.,
D.M.C. less 1 Div.
On 30th April, 60th Div.
attacked Shunet Nimrin
while the Aus. Mtd. Div.
captured Salt.

MAP TO ILLUSTRATE RAIDS TO SALT (APRIL 30TH, 1918) AND AMMAN
(MARCH 30TH, 1918).

III

MAP 9.

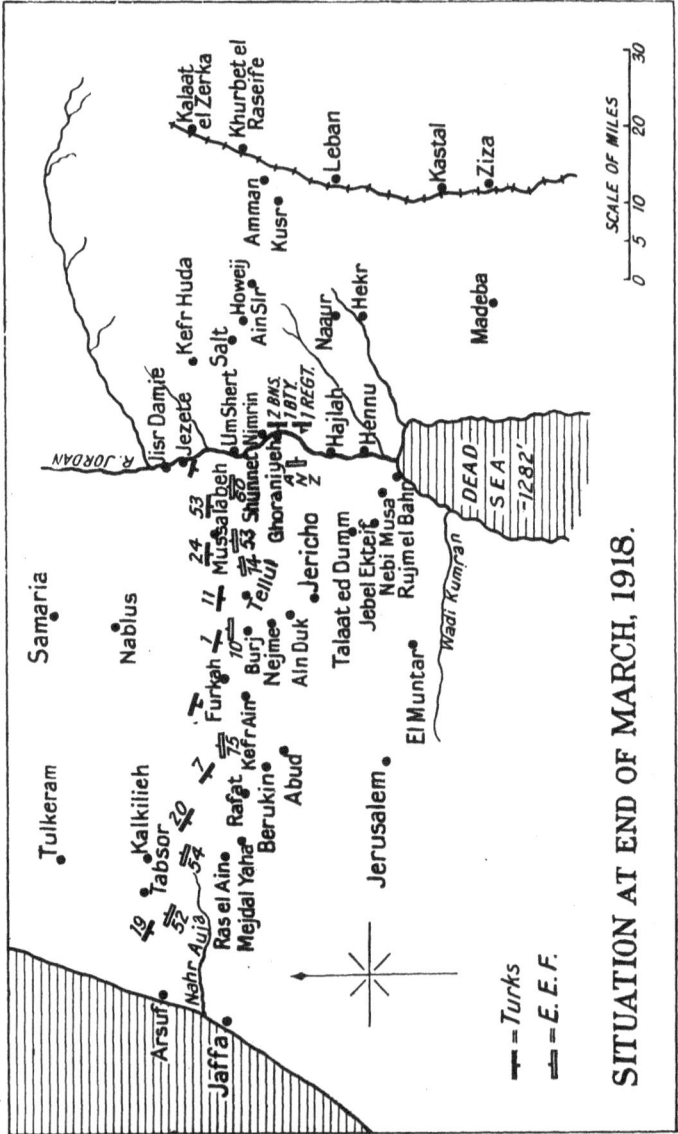

SITUATION AT END OF MARCH, 1918.

▬▬ = Turks
⊥⊥ = E.E.F.

II2

Map 10.

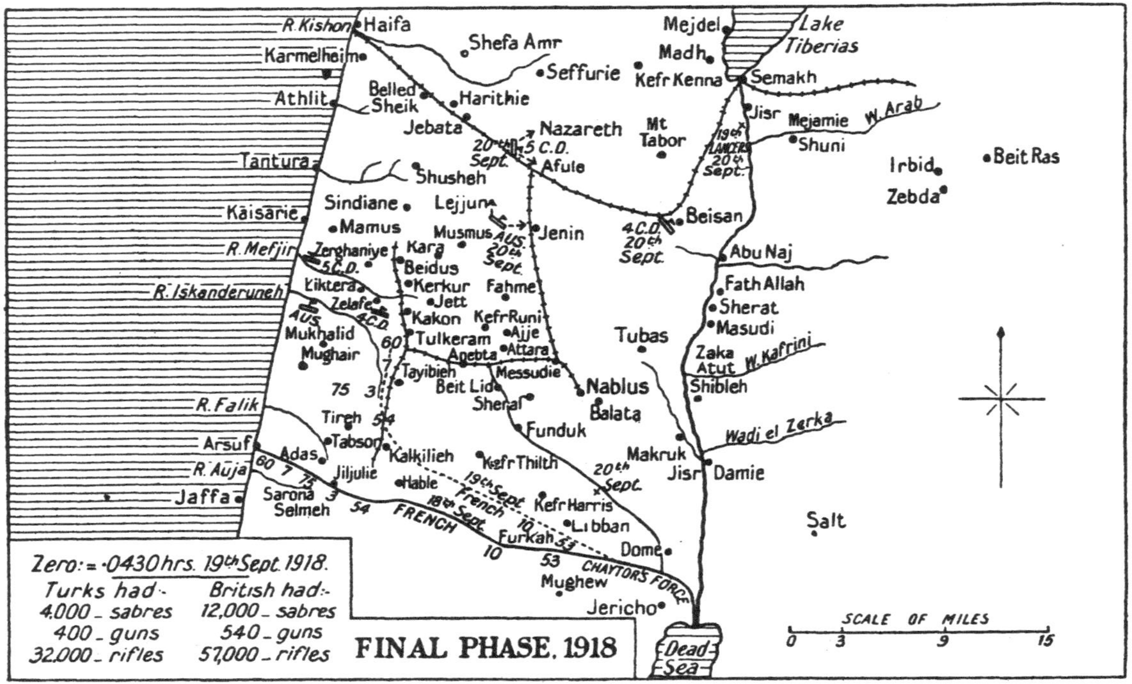

FINAL PHASE, 1918
Zero: = ·0430 hrs. 19th Sept. 1918.
Turks had:- 4,000 - sabres 400 - guns 32,000 - rifles
British had:- 12,000 - sabres 540 - guns 57,000 - rifles
SCALE OF MILES
0 3 9 15
R. Kishon
Haifa
Karmelheim
Athlit
Tantura
Kaisarie
R. Mefjir
R. Iskanderuneh
R. Falik
Arsuf
R. Auja
Jaffa
Shefa Amr
Seffurie
Belled Sheik
Harithie
Jebata
Nazareth
5 C.D.
20th Sept.
Afule
Shusheh
Sindiane
Lejjun
Mamus
Musmus
AUS. 20th Sept.
Jenin
Zerghaniye
5 C.D.
Kara
Beidus
Kiktera
Kerkur
Fahme
Jett
Zelafe
4 C.D.
AUS.
Kakon
Kefr Runi
Ajje
Mukhalid
Tulkeram
Attara
Mughair
Anebta
Messudie
Tayibieh
Beit Lid
Nablus
Balata
Sheral
Funduk
Tireh
Tabsor
Adas
Kalkilieh
Kefr Thilth
Jiljulie
Hable
Sarona
Selmeh
19th Sept. French
18th Sept.
FRENCH
Kefr Harris
Libban
Furkah
Dome
Mughew
Jericho
CHAYTOR'S FORCE
Dead Sea
Mejdel
Madh
Kefr Kenna
Lake Tiberias
Semakh
Jisr Mejamie
W. Arab
Shuni
19th LANCERS 20th Sept.
Mt Tabor
Beisan
4 C.D. 20th Sept.
Abu Naj
Fath Allah
Sherat
Masudi
Tubas
Zaka
Atut
W. Kafrini
Shibleh
Wadi el Zerka
Makruk
20th Sept.
Jisr Damie
Irbid
Zebda
Beit Ras
Salt

MAP II.

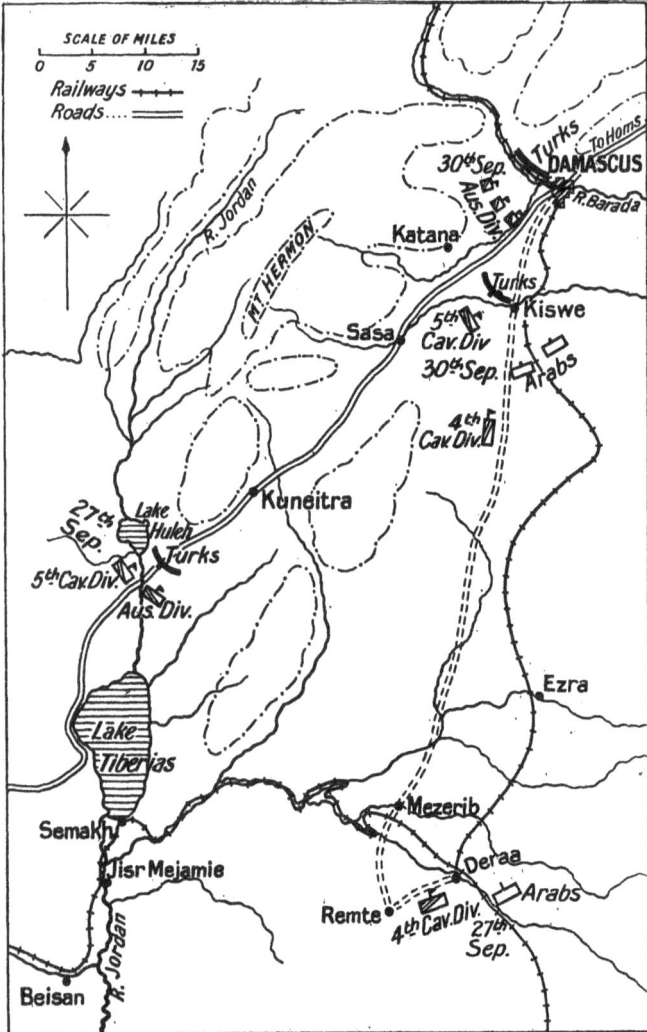

OPERATIONS FOR THE CAPTURE OF DAMASCUS.

MAP 12.

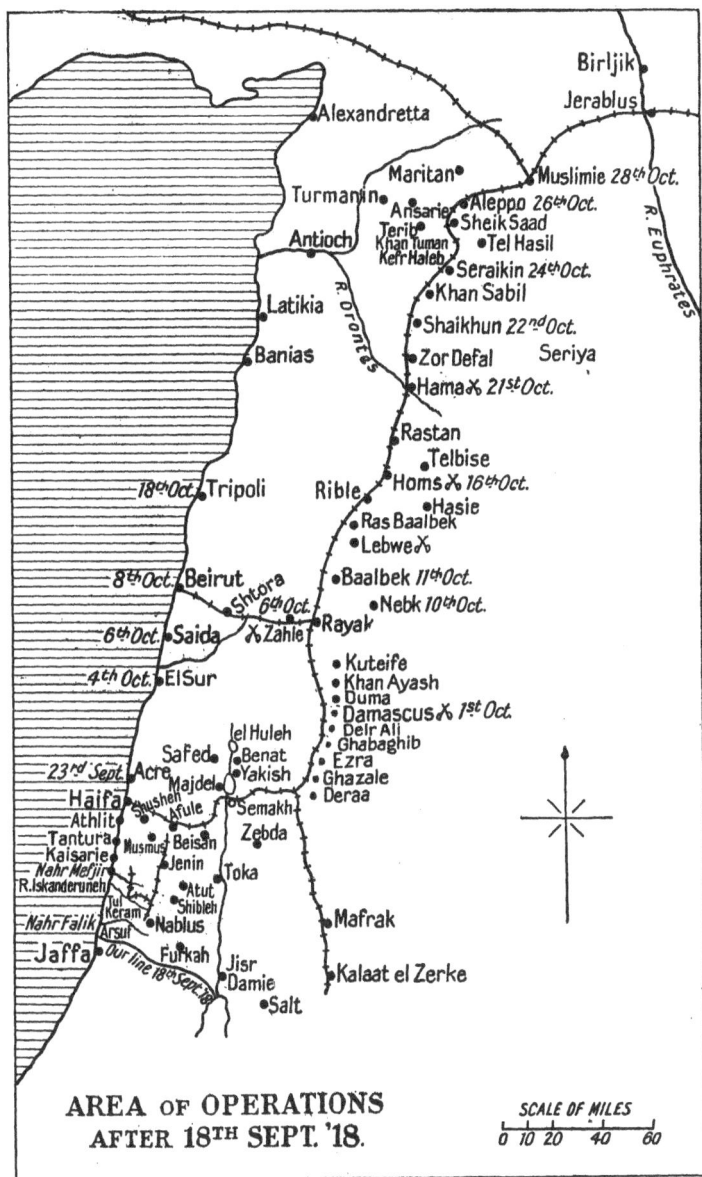

Birljik

Jerablus

Alexandretta

Maritan

Muslimie 28th Oct.

Turmanin

Ansarie

Aleppo 26th Oct.

Terib

Sheik Saad

Antioch

Khan Tuman

Tel Hasil

Kefr Haleb

Seraikin 24th Oct.

Khan Sabil

Latikia

Shaikhun 22nd Oct.

Banias

Zor Defal

Seriya

Hama ✗ 21st Oct.

Rastan

Telbise

18th Oct. Tripoli

Rible

Homs ✗ 16th Oct.

Hasie

Ras Baalbek

Lebwe ✗

8th Oct. Beirut

Shtora

Baalbek 11th Oct.

6th Oct.

Nebk 10th Oct.

6th Oct. Saida

✗ Zahle

Rayak

4th Oct. El Sur

Kuteife

Khan Ayash

Duma

Damascus ✗ 1st Oct.

Deir Ali

Iel Huleh

Ghabaghib

Safed

Benat

Ezra

23rd Sept. Acre

Majdel

Yakish

Ghazale

Haifa

Bushen

Afule

Semakh

Deraa

Athlit

Zebda

Tantura

Musmus

Beisan

Kaisarie

Jenin

Nahr Mefjir

Toka

R. Iskanderuneh

Atut

Shibleh

Tul Keram

Mafrak

Jaffa

Arsuf

Nablus

Furkah

Jisr

Damie

Kalaat el Zerke

Our line 18th Sept.

Salt

R. Orontes

R. Euphrates

AREA OF OPERATIONS
AFTER 18TH SEPT. '18.

SCALE OF MILES

0 10 20   40   60